ƒP

HOW AND WHY

YOUR

CUSTOMERS

ADOPT

TECHNOLOGY

Techno-Ready Marketing

A. Parasuraman

AND

Charles L. Colby

www.technoreadymarketing.com

THE FREE PRESS

New York London Toronto Sydney Singapore

The authors gratefully acknowledge permission from the following sources to reprint material in their control: B & O Railroad Museum, Inc., for the black-and-white photographic image of the painting "Race—Tom Thumb vs. Horse Car." U.S. Department of the Interior, National Park Service, Edison National Historic Site, for a photograph of Thomas Edison.

THE FREE PRESS
A Division of Simon & Schuster, Inc.
1230 Avenue of the Americas
New York, NY 10020
Copyright © 2001 by A. Parasuraman and Charles L. Colby
Designed by Brooke Zimmer Koven
Manufactured in the United States of America

10 9 8 7 6 5 4 3 2 1

Library of Congress Cataloging-in-Publication Data

Parasuraman, A.
    Techno-ready marketing : how and why your customers
adopt technology / A. Parasuraman and Charles L. Colby
        p.   cm.
    Includes index.
    1. Technological innovations—Management.
    2. Technological innovations—Marketing. I. Colby,
Charles L. II. Title.

HD45 .P327 2001
658.8—dc21   00-065457

ISBN 13: 978-1-4165-7663-1   ISBN 10: 1-4165-7663-0

To my brother-in-law, Balu, his wife, Raji, and their son, Hari, for their many acts of kindness.

—A. PARASURAMAN

To my wife, Debbie, and children, Sarah and Matthew, who provided encouragement and patience while we embarked on this project, and to my mother and fellow author, Petrina Aubol, who offered moral support and advice.

—CHARLES L. COLBY

Acknowledgments

More than a book, *Techno-Ready Marketing* is a major multi-year initiative involving research projects, case studies, presentations, and discussions with techno-ready marketers. Many organizations and individuals contributed directly or indirectly to our efforts by providing resources, advice, and insight. Scores of individuals assisted us, but we would like to at least acknowledge the organizations where they work.

The staff at Rockbridge Associates, Inc., helped us shape our ideas and implement the research studies on which our theories are based. Sallie Mae provided support for a baseline study of technology readiness. America Online, Freddie Mac, Interealty Corporation, and the Community College of Philadelphia supported our earliest efforts by providing opportunities for case studies. Staff at Bell Atlantic (now Verizon) Yellow Pages offered special encouragement and advice. A number of other corporations and associations gave us opportunities for testing our approaches or giving encouragement and business insights, among them, Amway Corporation, Best Software, Capital One, Carl M. Freeman Associates, Careerbuilder.com, Cox Interactive Media, Discovery Communications, Inc., GTE, Marriott International, MCI WorldCom, and PBS, as well as the Association for Supervision and Curriculum Development

(ASCD), Meeting Professionals International, and the National Rural Electric Cooperative Association.

Some associations were especially supportive in our efforts. The Consumer Electronics Association provided us valuable suggestions, access to their research library, and a forum for our ideas at the International Consumer Electronics Show (ICES 2000). The Cable and Telecommunications Association for Marketing (CTAM) was a strong supporter of our efforts and honored us with their research case study award in 1999. Many of the media have reported on our research, helping us disseminate our ideas, but some that were particularly supportive include the *San Diego Union Tribune, Future Banker,* and EcommerceTimes.com.

Our work on techno-ready marketing has also benefited from feedback and encouragement we received from colleagues during several academic conferences in which we presented findings from various phases of our research program. In particular, we would like to acknowledge the insights we gained by interacting with colleagues at the American Marketing Association's Frontiers in Services conferences and the International Service Quality Association's Quality in Services (QUIS) conferences. We would also like to acknowledge support from the University of Maryland's Center for E-Service, a cosponsor of the National Technology Readiness Survey, published by the authors.

CONTENTS

Preface   XI

PREFACE

Rapid advances in current technologies and the accelerating emergence of new ones are flooding the marketplace with innovative products and services. The marketing of technology-based products and services, however, is by and large being guided by traditional principles that may not be as effective for high-tech offerings as they are for their low-tech counterparts. Companies collectively possess a far bigger reservoir of technological savvy, the primary force behind the proliferation of innovations, than of the marketing savvy necessary for fully capitalizing on those innovations. There is a virtual vacuum of sound, research-based guidelines for effectively marketing innovations and leveraging technology to strengthen relationships with customers. This books attempts to fill that vacuum.

The two of us have been collaborating for over four years on a series of qualitative and empirical research projects, focusing on buyers' technology-related beliefs and behaviors, with each project verifying, clarifying, and building on the insights from its predecessor. In addition to our systematic program of scholarly research, we have been involved in dozens of consulting assignments pertaining to the marketing of technology-based products and services in a variety of sectors. The combined learning from our research and consulting experience compellingly suggests that consumer behaviors associated with cutting-edge technology and conventional offerings differ significantly. Therefore, companies wishing to reap maximum benefits from technology-based products and services must be cognizant of, and put into practice, the unique principles of techno-ready marketing—the subject matter of our book.

We use the term *techno-ready marketing* to capture the concepts critical for successfully marketing innovative products and services that are technology-intensive. We would argue that most companies are at best mediocre techno-ready marketers—in terms of not only marketing technological innovations but also harnessing technologies to foster customer-company interactions and gain competitive advantage. The primary reason underlying such techno-marketing mediocrity is inadequate understanding of customers' attitudes toward technology and important variations in those attitudes across different customer segments. Companies relying solely on conventional approaches to market technology-based offerings can fall prey to costly pitfalls. Moreover, such companies may be oblivious to opportunities for boosting their bottom lines by engineering faster acceptance of their market offerings and consolidating their competitive advantage.

In this book, we invoke insights from our extensive research and consulting experience, as well as from the business, historical, and social science literatures, to develop and discuss guidelines—in the form of principles, frameworks, and strategies—for excelling at marketing technology-based products and services. A key construct that emerged from our research and serves as the foundation to which the guidelines we propose are anchored is people's *technology readiness* (TR).

Representing an amalgam of feelings, hopes, fears, and frustrations about technology, the TR construct captures people's overall propensity to embrace and use new technologies for accomplishing goals in home life and at work. It is much more a mental state than a measure of technical competency. Our research consistently shows that technology readiness is *multifaceted,* and that the facets combine in complex ways to produce distinct types of technology-ready individuals, with differing behavioral processes pertaining to the adoption of technology-based products and services. For instance, consumers who are high on *innovativeness* (one of the TR facets uncovered by our research) may not necessarily be "technology ready" if they also experience a great deal of *discomfort* with technology (another TR facet).

This book discusses in detail the meaning and measurement of technology readiness, explores in depth an empirically derived *typology of individuals* based on their distinct patterns of TR scores, articulates the typology's relevance for understanding technology-related consumer behavior, and demonstrates how managers can use the typology to formulate marketing strategies for successfully acquiring and retaining technology customers. Throughout the book we have peppered our discussion of the various techno-ready marketing principles, frameworks, and strategies with supporting evidence from our research, as well as with numerous examples and case studies.

The insights this book offers are relevant for organizations of various sizes and types that currently are facing (or will soon face) technology-related issues in marketing to and/or interacting with customers. These customers may include consumers, businesses, or a combination of both. Moreover, the insights should be of interest not only to marketing executives but also to senior management and to decision makers involved with product/service design, research and development, human resources, training, information systems, and operations. We have worked hard to keep the presentation succinct and accessible to a broad spectrum of executives. We hope you enjoy reading the book, find its contents thought-provoking and inspiring, and learn something about increasing the techno-readiness of your organization.

A. PARASURAMAN
CHARLES L. COLBY

Techno-Ready Marketing

CHAPTER 1

The Craft of Techno-Ready Marketing

In an environment transformed by technological change, the best technology solution does not guarantee market success. The career of Thomas Alva Edison is a case in point. With over 1,000 patents to his name, he was one of the most prolific inventors in history. Edison possessed an uncanny ability to identify a lucrative opportunity and respond by commercializing a new technology with broad market appeal.[1]

Throughout his 60-year career, which ended only upon his death in 1931, Edison saw potential in a wide range of products that included improved telegraph transmission, the phonograph, electric lighting, construction cement, motion pictures, electric cars, and rubber refining. In almost every case, his efforts led to stunning success in the early formation of a market he helped bring into being. In the long run, his ventures were usually surpassed by ambitious competitors who were better able to navigate the turbulent market for the new technology. His company, TAE Enterprises, never achieved the status of an AT&T, IBM, or Microsoft.[2]

One particularly instructive example is the phonograph, which Edison first commercialized in the 1870s. Successful as a novelty item with a limited market, the invention languished for decades until the efforts of budding competitors caught Edison's attention. One of these competitors was the Victor Talking Machine Company, which today might be called a "technology start-up."

The Edison phonograph technology involved a rotating cylinder upon which sounds were etched by a recording needle. In 1908, Victor introduced a rotating disk (the record format used for decades until made obsolete by CDs in the 1980s). Edison clung to the cylinder technology because he believed it provided a higher quality reproduction of sound due to the constant speed of rotation. But the Victor disk system was capable of storing a longer playing recording. Consumers did not notice the difference in sound quality, but they did see an advantage to more music on an individual storage unit.

Under great pressure from distributors, Edison reluctantly authorized the introduction of a disk system in 1912. In an effort to improve on the technology, his company developed a thicker, heavier disk made of compressed wood flour covered with varnish. He argued that his disks, which weighed 10 ounces, would be less likely

EXHIBIT 1-1

One of the most prolific inventors in history, Thomas Edison might have achieved greater commercial success if he had been a more savvy techno-ready marketer who was sensitive to customer requirements. He is shown here with an early version of a phonograph that used a rotating cylinder to record and store sound. *Source: Edison National Historic Site, National Park Service, West Orange, N.J.*

to shatter than the frailer disks made by the competition. But this feature offered little advantage because consumers were unlikely to drop their precious disks.[3]

Edison tried other strategies to gain an edge in the

market. In a decision ominously similar to one made decades later by Sony with its Betamax VCR format, Edison created his own standard. He deliberately introduced disks that were incompatible with the Victor system (Edison's disk rotated 80 times per minute, compared to Victor's 78½). Edison then sought to monopolize the best recording talent exclusively for his system, personally selecting and screening the artists. This plan had one problem: the great inventor suffered from a serious hearing impairment, so his taste in music was rather idiosyncratic and not attuned to those of the general public.[4]

Victor proved to be a much more savvy technology marketer. For example, it signed up its own popular artists of the time, including Enrico Caruso of the Metropolitan Opera. The company also moved into technologies that met new consumer needs, such as electronic recording and radio. In 1929, facing mounting losses and unable to survive solely on sales in the lucrative high-end market where it had gotten its start, the Edison family closed down its phonograph business.[5]

PRINCIPLES OF SUCCESSFUL TECHNO-READY MARKETING

Although he was not the most successful marketer, Thomas Edison claims an important place in history. Edison's craft can be labeled *techno-ready marketing*, the process of creating and developing markets by deploying innovative technologies. The word *techno* implies advancing a body of knowledge; the term *innovation*, often used in conjunction with technology, implies striving for something new, fresh, novel, and unexpected.[6] Used in reference to a product or service, an innovation is unique because it is cutting edge when it is introduced. Examples of technology innovations today include e-commerce (selling on the Internet) and genetically enhanced food crops. In most cases, technology-driven innovations remove a certain degree of human input from the creation and delivery of a product or service. Well-known cases include the introduction of the automated teller machine (ATM) replacing a bank teller, and a typewriter replacing a scribe.

Techno-ready marketing is the science and practice of marketing products and services that are innovative and technology-intensive, a subject of great interest because our times are shaped by an explosion in innovation. Techno-ready marketing should be considered a separate discipline within the broader science of marketing because of the unique critical success factors when technology is involved. Furthermore, the factors that lead to satisfactory customer relationships are different for technology.

The uniqueness of techno-ready marketing can be characterized by four core principles about technology markets:

- *Principle 1: Technology adoption is a distinct process.* Technology marketers are successful only if their output is accepted. The customer behavior for a technology-based product or service differs from a more conventional one. We have discovered this through extensive research on consumer beliefs about technology. When the marketer is introducing a cutting-edge product that replaces more of the human element, a whole set of special consumer beliefs comes into play. This includes a varying level of optimism about technology, a tendency to innovate, a problem of discomfort with technology, and an inherent insecurity. These beliefs are less germane, if at all, when marketing a tastier cereal, a classier vehicle, or a smoother-shaving razor.

- *Principle 2: Technology innovations require different marketing strategies.* Because the adoption process is different when technology is involved, so must be the approach to product design, pricing, communication, distribution, and service. To illustrate, a soft drink (a product with little technology) might best be marketed by using advertising to build a positive image in the minds of the target audience, one that appeals to the desired self-image of the buyer. A computer also affects a consumer's self-image, but primarily by how *user-friendly* it is. For instance, if the computer fails to operate as desired in front of others, the user could be embarrassed and suffer a loss of self-esteem. As such, a computer maker would do well to initially aim its advertising more selectively at early adopters who are more confident about using technology. The computer maker must also channel considerable effort into making the product easy to use. Once the product is in the market, the computer maker must direct efforts toward helping new users operate it.

- *Principle 3: Ensuring customer satisfaction is a more weighty challenge for a technology-based product or service.*

Once consumers adopt, they must grapple with an unfamiliar and often more complex approach to satisfying their needs. Customers of technology-based offerings require education and support. Furthermore, customers will vary considerably in the level of help they require and in their receptivity to the support that is offered. To illustrate, a customer who walks into a bank branch will engage mostly in familiar transactions and will have access to employees to solve problems. A customer who starts to bank online will need to learn banking anew and will be faced with uncertainties about the entire process.

• *Principle 4: Technology markets are governed by a law of critical mass, often resulting in a "winner takes all" outcome.* In a technology-driven market, it is not uncommon for a single company to achieve a dominant position that, once achieved, is impossible to challenge until a whole new technology comes along. Early entrants offering a new technology can be quite successful, but ultimately, one company engulfs its competitors or relegates them to niche status. History has shown that this occurs as a result of economies of scale in production, ownership of a standard, or interconnectivity. Examples of techno-ready marketers who achieved dominance these ways include, respectively, the Ford Motor Company with its mass production of the Model T, Microsoft with its standardized Windows operating system, and AT&T with its telecommunications network, which remained a monopoly until the 1980s.

EXHIBIT 1–2

THE CYCLE OF CUSTOMER-FOCUSED INNOVATION

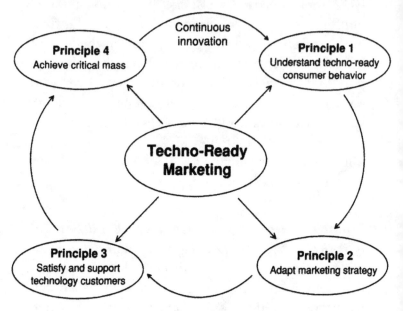

As shown in Exhibit 1–2, each principle corresponds to a marketing practice, beginning with an effort to understand techno-ready consumer behavior. The logic of the first principle has been verified through consumer research, including work conducted by the authors. The first principle supports the logic of subsequent principles: if consumer behavior is indeed unique, then it is logical that marketing, servicing, and the path to success should also be unique. Eventually, an innovation ceases to be such, and a techno-ready marketer is left with the choice of competing in a mature market or innovating with technology as part of a continuous process.

EXHIBIT 1-3

RATE OF ADOPTION OF A COMPANY'S
TECHNOLOGY-BASED PRODUCT OR SERVICE

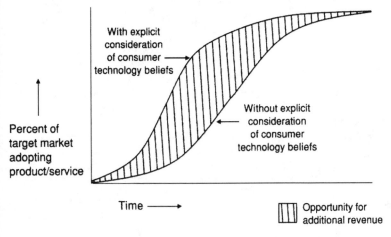

Many organizations engaged in techno-ready marketing could easily ignore the principles described
above. Like Thomas Edison in the early years of marketing a hot new innovation like electric lighting, they
could argue that their product resulted in a sustained
commercial success. They could do more to understand
consumer beliefs and adapt accordingly, but who can
argue with profits? Marketers who hold this viewpoint
should consider a few caveats relating to the principles
above.

- If a company markets its technology in a manner
  that is more responsive to customer need, it will
  accelerate the speed of adoption and achieve greater
  sales volume sooner. (See Exhibit 1-3.)

EXHIBIT 1-4

TECH-SUPPORT COSTS OVER A TECHNOLOGY LIFE CYCLE

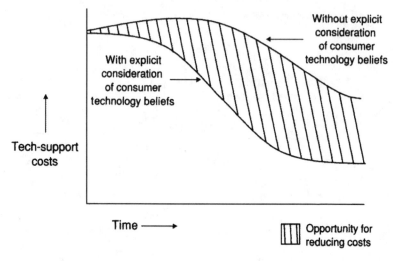

- Poorly designed technology results in products and services that are harder to use. This taxes the customer support infrastructure, costing money. Technology that is *customer-focused* will cut product returns and service cancellations. It will also result in fewer calls to tech support. All of these outcomes reflect positively on the bottom line. (See Exhibit 1-4.)

- Perhaps the most pressing reason for being a smart techno-ready marketer is the principle of critical mass. It is possible for one company to completely win the game. The winner will be the company that does the best job of responding to the unique beliefs, adoption process, and educational requirements surrounding a technology-based product or service.

TALE OF TWO SERVICES: HOW BEING THE "FIRST" DOES NOT GUARANTEE LEADERSHIP FOREVER

Close to half of U.S. households use an online or Internet service, and growth in this market shows no signs of slowing. If credit were to be given to one company for creating this huge industry, a good candidate would be the venerable pioneer CompuServe. Before the World Wide Web even existed and few people had heard of the Internet, CompuServe had over a million subscribers interconnected by computer. The company was acquired in 1993 by H&R Block, and was considered at the time to be the crown jewel in the company's portfolio.

How did this innovative company become so successful? To begin with, CompuServe offered a truly unique service that satisfied an unfulfilled niche. It used network technology to provide a mass market with instant access to valuable databases. It also connected users with each other, allowing them to interact by e-mail and forums. Another key to CompuServe's success was its focus on *serious* users—the business, professional, and technical customers—who were the only ones willing to pay for the information-rich service.

Compared to the Internet world of today, the CompuServe system was a challenge to use. It required some tech-savviness to install, its IDs were long streams of numbers instead of an actual name, and file transfers over the Internet required extra steps. These obstacles did not stop the early consumers who were willing to innovate and could see the benefits of being online.

EXHIBIT 1–5

AMERICA ONLINE SUBSCRIBERS

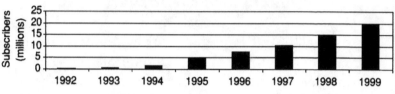

Source: AOL Communications

While CompuServe was growing a base of loyal users, a technology entrepreneur named Steve Case was pursuing a vision of his own. He wanted to introduce online computing to a mass market. In his company, which took the name America Online in 1985, the key word would be simplicity, and the customer an average person who could be totally lacking in technology skill. The company got off to a rough start. Case was almost fired by his board after the first year, but a board member argued that the company could benefit from the lessons he gained from losing millions of dollars. Ultimately, the company began to accumulate market share rapidly—one million subscribers in 1994, seven million in 1996, and 10 million in 1997,[7] when America Online announced that it was acquiring the floundering CompuServe.

What was the secret to America Online's growth? Technology pundits will argue that many factors contributed to this success, including aggressive mass marketing, attractive pricing, and its frequently visited chat rooms. But AOL offered a singular benefit that allowed it to accumulate subscribers at an explosive pace: *ease of use.*

To begin with, the service was easy to install; subscribers needed to do no more than pop a start-up disk in their computers and enter a credit card number. Second, it was easy to operate, relying on a neatly organized screen to help the user navigate through the range of information options. Third, it made online communication a simple and joyful process; screen names could be chosen by the customer, e-mail was a menu-driven process, and users could visit the popular chat rooms without having to load special software.

It is popular among technology elitists to condescendingly categorize America Online as a service for the novice, so when it finally overtook CompuServe in number of subscribers in 1995, experts gleefully predicted the eventual demise of AOL, which was already experiencing a high rate of customer turnover. They pointed out that AOL customers would most certainly gravitate to more serious services that gave them more direct access to the Internet.

The most interesting part of the AOL story is not how it got started but how it achieved its position today as the dominant provider of consumer Internet services. Its market share is 45% in a highly fragmented consumer market.[8] No company comes close to challenging its position, and for reasons to be discussed, this position is unlikely to change in the near future. AOL went through a turbulent period in 1996 when it became overwhelmed by new subscribers and encountered extensive negative publicity about its service quality. Research we conducted at the time, however, showed that customers were grumpy but forgiving. The Internet was still a wild frontier where nearly all online alternatives created frustrations for the first-time user. AOL at least got their customers online. It maintained its simplicity of use and eventually corrected its quality problems, making it even easier to use. It aggressively seeded the market with free start-up disks and arranged to be a common choice for automatic installation on new computer equipment.

AOL now controls one of the most valuable pieces of real estate, its portal to the online world. No other Internet service provider can match the huge volume of traffic passing through their site on the way to the Internet, or the tens of millions who rely on the service to direct them to their desired locations. As an indication of the value this offers, Tel-Save, a long-distance marketer, paid AOL $100 million for exclusive access to subscribers for three years. Other deals abound, such as a $50 million deal to carry the online shopping service of CUC International, and a $32 million deal with Preview Travel.

Now that it has a dominant position, no competitor can extract such lucrative deals. With healthy revenue sources besides subscriber fees, AOL has the ability to keep prices within reason, market its brand aggressively, and maintain its dominant position. AOL has mastered the fourth principle of successful techno-ready marketing. It has achieved critical mass.

A MAP OF THIS BOOK

This book covers two broad areas: (1) research-based observations about the process consumers undertake in purchasing technology and (2) implications for succeeding as a techno-ready marketer by taking advantage of this knowledge. The next four chapters provide a detailed understanding of behavior and how it operates in marketplaces, based on years of rigorous research conducted by the authors. Specifically, Chapter 2 introduces the concept of technology readiness, an important construct that explains how people react to technology innovations. Chapter 3 is a more scholarly treatment of this construct, identifying the detailed components of technology beliefs. Chapters 4 and 5 present a consumer typology, or segmentation, that classifies people by their different technology beliefs and shows the role each plays in the ascent of a new technology.

The remaining chapters discuss how organizations can succeed in the practice of techno-ready marketing, providing case studies from history and modern business. The discussion begins (in Chapter 6) with the introduction of the pyramid model of marketing, depicting how technology influences the marketing practices of organizations. The next chapter (Chapter 7) addresses techno-ready marketing more specifically, presenting practices for acquiring customers and growing market share. The chapter that follows (Chapter 8) considers how to satisfy customers, including implications for design, customer service, and communications. These ideas are assimilated in the next chapter (Chapter 9) on the Techno-Ready Marketing Audit, which helps readers assess their own organization to determine what steps to take to become a more successful techno-ready marketer.

Although this book focuses on techno-ready marketing in business, we believe it offers valuable insight for professionals in academia, government, and the non-profit sector. The final chapter (Chapter 10) concludes with a discussion of the techno-ready society and examines the broader implications of technology readiness in the world of the future.

CHAPTER 2

Technology Readiness

I f a techno-ready marketer desired a single revelation that would lead to spectacular success, it could well be the concept of technology readiness, or TR, which describes the distinctive behavioral process behind the adoption of technology-based products and services. TR is the gist of the first principle of successful techno-ready marketing: that customer beliefs and behaviors correspond to a different model when cutting-edge technology is involved. This chapter focuses a microscope on the technology consumer, showing how the decision process is unique and distinct from that of conventional products.

A CUTTING-EDGE PHENOMENON WITH ANCIENT ORIGINS

Technology readiness refers to people's propensity to embrace and use new technologies for accomplishing goals in home life and at work. The next chapter discusses in detail all the important components of TR, but a few things should be noted for now.

- **TR varies from one individual to the next.** Anyone can be a consumer of a technology, but the path to adoption and the implications for marketing will depend on the degree and nature of the individual's TR. Some consumers may actively seek technology, while others may need special help or coaxing.

- **TR is multifacted.** It is more than just a tendency to be an "innovator" or "early adopter." Different types of beliefs blend to produce one's overall TR.

- **TR predicts and explains consumer response to new technologies.** It predicts the adoption rate of new technologies and also explains the manner in which they are used. Furthermore, it is associated with the degree of satisfaction with technology and the kinds of support required.

We have invested over three years studying TR and its managerial implications. Even prior to our formal research, there has been ample evidence that the public has always viewed technology in a unique context. This is illustrated by the legend of the Tom Thumb, a loco-

The race of the Tom Thumb steam engine, a true historical event, is also an example of technology as a critical part of American folklore.

motive that raced a horse-drawn train in Baltimore in 1830. The race was a true historical event, but the story has also become an important part of American folklore. In the telling of the tale, the archaic steam engine and its inventor, Peter Cooper, are the protagonists, and technology is given a positive spin. The story ends with the Tom Thumb losing to the horses when an engine belt slips. Anyone privy to the story knows that the ending is ironic, since the steam locomotive eventually dominates and transforms people's lives in a dramatic fashion.[1]

The Tom Thumb tale is one of many dealing with

the interplay between people and technology, all showing that technology has historically been salient in people's minds. Another example is the story of John
Henry, a labor hero who defeats a steam drill in a race
that confirms the worthiness of human endeavor when
threatened by the relentless advance of automation.
Clearly, technology's benefits and threats have always
occupied center stage in the collective minds of
societies.

The theme of technology influencing our lives has
been more than a subject of legend. It is a reality that
has shaped society throughout time. The following
examples illustrate:

- Throughout history, advances in materials have
  transformed economies and lifestyles. The shift from
  flint to bone led to the creation of the barb, which
  greatly improved fishing and, in turn, the livelihood
  and diet of humans. The development of metal man-
  ufacturing resulted in tools that did not chip and
  could be hammered back into shape when they
  warped, which resulted in the construction of houses
  from felled timber, to name one of the countless
  changes on societies.[2]

- Printing technology freed societies from reliance on
  knowledge provided orally or through hand-
  prepared manuscripts. This technology advanced
  science by offering more reliable and efficient distribution of information.[3]

- In Europe in the Middle Ages, the stirrup changed
  the science of military defense by solving the simple

problem of helping a rider remain on a horse. It ended up playing a significant role in the creation of the feudal system.

- The safety elevator, invented by Elisha Otis in the 1850s, provided a high-speed method of lift that was free of accidental collapse. This innovation paved the way for construction of skyscrapers, which have left their mark on modern cities.[4]

- While many would cite the late twentieth century as the period of an unprecedented information revolution, similar advancements occurred in other ages. For example, Thomas Edison's first commercial successes were in advanced telegraph systems. While a growing technology-based service in the 1990s is online investing, the investment industry also proved to be one of the most avid users of cutting-edge telegraphy in the nineteenth century. Then, as now, timing was critical. The telegraph brought real-time information from stock exchanges into brokerage offices.[5]

The point made here is that the influence of technology is ancient, its role in our lives archetypal. People will be grappling with technology challenges and opportunities long after the World Wide Web is as mundane as the automobile or radio.

Today, the influence of technology readiness is evident in many emerging technologies. TR is a force behind the sudden exponential growth in portable telephones and Internet access and is a factor behind the protest movements attempting to block the introduction of genetically engineered crops. Its influence is particularly evident in the area of e-commerce, a case worth exploring in more detail.

Fear and Confusion: The Case of E-Commerce

The National Technology Readiness Survey (NTRS), an exhaustive survey of technology beliefs in the United States, conducted by the authors, reveals the dynamics behind the development of e-commerce.[6] The survey includes many topics that help define and quantify TR and its relationship to technology behaviors.

According to the NTRS, e-commerce is a widespread activity, with well over a third[7] of consumers who are online at home indicating they engage in some kind of shopping. There is also extensive commercial activity in services, such as online banking and travel. The future of e-commerce in this formative stage is unclear. There is considerable evidence of rapid growth, but also signs of great consumer reluctance. The NTRS sheds more light on this.

- Most e-commerce activity consists of purchases under $100. Consumers express a high degree of interest in purchasing small items over the Internet, such as tickets to events (46% consider this to be desirable).

- Consumers show a great reluctance to purchase large items, such as a car or furniture (only 14% consider this desirable). They are also concerned with making major financial commitments, such as a home mortgage, over the Internet.

- There is a huge gap in activity based on the level of technology readiness of the online consumer. Consumers can be divided into those who are "low," "medium," or "high" in TR. The "high-TR" online consumer is more than twice as likely to have shopped online than someone who is only "low" or "medium" in technology readiness.

A critical issue confronting marketers planning to launch a business on the Internet is: What are consumers thinking when confronted with the option of buying online versus doing business in the usual manner? Some insights are offered by the NTRS:

- Most consumers believe in the benefits of a technology such as e-commerce. Almost two-thirds (63%) like the idea of doing business via computers because they are not limited to regular business hours. In addition, 58% believe technology gives them more control over their daily lives. A sizable number of consumers (but a minority) think that the use of a computer in a transaction frees them from sales pressure, and that computers are easier to deal with than people.

- Despite the benefits, consumers are skeptical of displacing people in a transaction. The vast majority, 90%, think the "human touch" is important when doing business with a company. Consumers also possess a degree of skepticism about using technology for new areas like e-commerce. Two-thirds (64%) believe the benefits of new technology are often "grossly overstated."

- Perhaps the greatest obstacle to e-commerce today, particularly for expanding into areas where greater commitment is required, is related to perceived security and the need for assurance about the transaction. For example, 77% of consumers do not consider it safe giving out a credit card number over a computer, and 67% do not feel confident doing business with a place that can only be reached online.

The NTRS highlights the unique challenges that marketers face when dealing with advanced technology. A traditional retailer opening a new store does not have to grapple with emotional themes such as the value of human interaction, the fear of the unknown, or the need for assurance. On the other hand, before opening a virtual store on the World Wide Web, the retailer must carefully assess customers' technology beliefs. For example, one might conclude that a staid brand name such as Sears Roebuck, Procter & Gamble or United Parcel Service would be an anachronism in the world of e-commerce, but this may be far from the truth. With the introduction of a technology-based medium, consumers become insecure, making brand familiarity a more critical asset.

All the beliefs just described that influence the acceptance of e-commerce define different components of technology readiness. As with any kind of market

EXHIBIT 2–2

WHAT'S YOUR TR INDEX?

AN ABBREVIATED TECHNOLOGY READINESS SCALE

Directions: Indicate whether you "strongly agree," "somewhat agree," are "neutral," "somewhat disagree," or "strongly disagree" with the following statements:

Strongly Agree 5	Somewhat Agree 4	Neutral 3	Somewhat Disagree 2	Strongly Disagree 1

a. I can usually figure out new hi-tech products and services without help from others.
b. New technology is often too complicated to be useful.
c. I like the idea of doing business via computers because you are not limited to regular business hours.
d. When I get technical support from a provider of a high-tech product or service, I sometimes feel as if I'm being taken advantage of by someone who knows more than I do.
e. Technology gives people more control over their daily lives.
f. I do not consider it safe giving out a credit card number over a computer.
g. In general, I am among the first in my circle of friends to acquire new technology when it appears.
h. I do not feel confident doing business with a place that can only be reached online.
i. Technology makes me more efficient in my occupation.
j. If you provide information to a machine or over the Internet, you can never be sure if it really gets to the right place.

research, TR beliefs can be collected from a target market, analyzed, and translated into strategy. The second half of this book deals with the strategic implications of this new consumer theory.

WHAT IS YOUR TR INDEX?

Readers should try to evaluate their own level of technology readiness as they proceed through this book. Exhibit 2–2 presents a condensed version of the Tech-

EXHIBIT 2-2 (CONTINUED)

INTERPRETING YOUR RESPONSES TO THE TR INDEX		
Compute your Technology Readiness Index as follows: $(a + c + e + g + i) -$ $(b + d + f + h + j)$ If your index score is . . .	Your percentile among the adult U.S. general population is . . .	You would be considered . . .
16	99%	
14	98%	
12	97%	Highly techno-ready
10	94%	
8	91%	
6	86%	
4	79%	Somewhat techno-ready
2	70%	
1	65%	
0	59%	
-1	51%	Average
-2	44%	
-4	34%	
-6	24%	Somewhat techno-resistant
-8	19%	
-10	11%	
-12	7%	Highly techno-resistant
-14	5%	
-16	2%	

Copyright 1999 by A. Parasuraman and Rockbridge Associates, Inc.
Replication of this scale or its use for any commercial purpose requires written
permission from the authors.

nology Readiness Index that readers can administer to themselves. It is important to note that the index is not a test of competency. It serves as an insight into a person's own motivations and inhibitions regarding the adoption of technology, factors which are discussed in Chapter 3.

CHAPTER 3

The Technology Readiness Index (TRI)

People's technology readiness—their propensity to embrace and use new technologies for accomplishing goals in home life and at work—is an overall state of mind rather than a measure of competency. It is a combination of technology-related beliefs that collectively determine a person's predisposition to interact with technology-based products and services. Based on our multiyear research program focusing on people's technology readiness, we have developed a metric—the Technology Readiness Index (TRI)—for measuring it.[1]

In this chapter, we first offer a brief overview of the conceptual foundation in which the TRI is anchored. We then discuss and illustrate the multiple facets of the TRI.

EXHIBIT 3-1
NTRS METHODOLOGY

The National Technology Readiness Survey (NTRS) is a nationwide survey of American adults (18 years of age or older). The primary purpose of the NTRS is to provide an in-depth view of consumer beliefs about new technologies. In addition to offering comprehensive information about people's technology beliefs, the NTRS also examines:

- Consumers' technology vision
- Employees' technology vision
- Usage of technology-based products and services
- Impact of the Internet on behavior
- Cellular phone usage
- Desired methods of tech support
- Association between people's technology beliefs and their demographics and lifestyles

The NTRS questionnaire, containing some 190 questions, was administered via computer-assisted telephone interviews to 1,000 adults between December 1998 and January 1999. Households were selected using random-digit-dialing techniques and received a maximum of eight callbacks in order to obtain a completed interview. Gender quotas and sample weighting were used to ensure that the sample was representative of the U.S. population as defined by the 1998 Current Population Survey (U.S. Census Bureau). In order to ensure quality in data collection, interviews were monitored, and call-center supervisors recontacted a sample of respondents for verification purposes.

The NTRS is an annual survey that tracks changes over time. The 2000 NTRS focused on e-commerce as a case study. The University of Maryland's Center for E-Service is a cosponsor of the NTRS.

We conclude the chapter with findings from the National Technology Readiness Survey pertaining to the TRI's ability to distinguish between different levels of ownership and usage of technology-based products and services. (See Exhibit 3–1 for a description of the NTRS.)

CONCEPTUAL UNDERPINNINGS OF THE TRI

Insights from dozens of customer focus groups that we have conducted in a variety of sectors (e.g., financial services, online services, electronic commerce, telecommunications, software) consistently suggest that people can simultaneously harbor favorable and unfavorable beliefs about technology. Exhibit 3–2 presents quotes from our focus group research that illustrate the types of positive and negative feelings that consumers have toward technology.

While people's positive views may propel them toward technology, their negative views may tug them away from it. Findings from studies conducted by other researchers also support the coexistence of push and pull forces that together determine a person's propensity to embrace technology. For instance, based on a comprehensive study of people's reactions to technology, Professors David Mick and Susan Fournier identified eight "paradoxes" of technology with which consumers have to cope: control/chaos, freedom/enslavement, new/obsolete, competence/incompetence, efficiency/inefficiency, fulfills/creates needs, assimilation/isolation, engaging/disengaging.[2] As these paradoxes imply, technology may trigger both positive and negative feelings.

Consider the Internet, arguably the most pervasive and talked-about new technology to be introduced in recent times. While press reports frequently tout—and people appreciate—the personal, societal, and commercial benefits of the Internet, evidence of some paranoia pertaining to this revolutionary technology also exists. The following list illustrates the kinds of concerns people have about the ill effects of the Internet:

- Proliferation of pornography[3]

- Surveillance by "Big Brother"[4]

EXHIBIT 3-2

EXAMPLES OF POSITIVE AND NEGATIVE BELIEFS
ABOUT TECHNOLOGY

The following quotes are excerpted from focus group interviews with customers in a variety of sectors. The type of customer is shown in parentheses following each quote.

Positive Feelings about Technology

"I have electronic banking service so I don't have to write checks. It's awesome." (Borrower)

"More and more students are sending us e-mails and we're able to reply to them at all hours of the day and night." (Education professional)

"There are some practices where human contact can get in the way and banking is definitely one of them. I take advantage of automated phone systems and online banking." (Bank customer)

"Online you don't have to worry about rushing to the store before it closes and dealing with traffic." (Shopper)

"I think any online help capability is always useful, because it's much easier to click on an icon, rather than track down the manual." (Business software user)

"If I've got a problem on the computer, that's the first place I go—the Net; throw it up on a newsgroup and I get an answer back in about an hour." (Internet user)

"It had a lot of information available. It made you feel as if you were making an informed decision." (Traveler)

Negative Feelings about Technology

"You have paid for all this extra hardware, and what do you do with it?" (Internet user)

"It's been frustrating because every time you think you're up to date you look at somebody else's Web site and recognize that you've fallen another step behind." (Education professional)

"We're usually not the first people in electronically. We like to see other people stumble—see what happens." (Banker)

"There's always some glitch [with technology]." (Traveler)

"I am not that brave on the computer." (Internet user)

"I would have loved to see a confirmation e-mail almost immediately that wasn't just an automatic e-mail. There should be a name attached to it that says I am a real person and you are booked." (Traveler)

"For me, it's a safety factor. I like hearing a voice." (Traveler)

- Information overload[5]

- Unreliable information[6]

- Loss of human interaction[7]

- Increasing chasm between "haves" and "have nots"[8]

- Loss of national and cultural identity[9]

- Fostering of criminal activities/terrorism[10]

These concerns are psychological barriers that dampen enthusiasm for embracing the Internet. Likewise, negative beliefs about new technology in general are likely to adversely affect people's technology readiness.

Although positive and negative feelings about technology may coexist, people are likely to vary regarding the relative dominance of the two types of feelings. In other words, we can array people along a "technology-beliefs continuum," anchored by "receptive" at one end and "resistant" at the other. Moreover, people's positions on this continuum can be expected to reflect their technology readiness (see Exhibit 3-3).

Findings from a handful of studies relating to consumer/technology interactions offer empirical support, albeit indirect, for the conceptual link between technology beliefs and technology readiness shown in Exhibit 3-3. For instance, several studies dealing with interactive media suggest the presence of distinct customer segments with differing beliefs about, and acceptance of,

EXHIBIT 3-3

LINKS BETWEEN TECHNOLOGY BELIEFS AND TECHNOLOGY READINESS

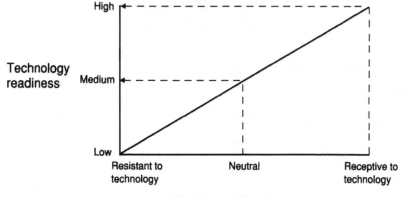

Technology-Beliefs Continuum

the media.[11] Likewise, a study of technology-based self-service options found that consumers differ in their beliefs/feelings about the various options, and that consumers with more positive beliefs/feelings are more likely to use the various options.[12]

As the foregoing discussion implies, a person's position on the technology-beliefs continuum is reflective of his or her degree of technology readiness. The TRI is a multiple-item scale that measures people's technology readiness by ascertaining their positions on 36 different technology-belief statements. Roughly half the statements in the TRI relate to positive beliefs (e.g., "Technology gives people more control over their daily lives"), while the rest relate to negative beliefs (e.g., "New technology is often too complicated to be useful"). The condensed, 10-item version of the TRI presented at the conclusion of Chapter 2 contains additional examples of the belief statements.[13] As in the case of the 10-item TRI, each of the 36 statements in the full TRI are scored on a 5-point scale (1 = *Strongly Disagree*, 2 = *Somewhat Disagree*, 3 = *Neutral*, 4 = *Somewhat Agree*, and 5 = *Strongly Agree*).

In constructing the 36-item TRI, we began with an initial version of the scale by drawing upon insights from our extensive qualitative research as well as from other relevant technology-related studies such as the ones mentioned earlier. We then undertook a multistage program of research that involved a series of empirical studies to strengthen and refine the scale, and verify its reliability and validity. Details about the various stages of this iterative scale-development process are available elsewhere.[14] We turn now to the structure and composition of technology readiness, as captured by the final TRI scale.

THE MULTIPLE FACETS OF TECHNOLOGY READINESS

The insights from our qualitative research suggesting that people's beliefs about technology have both positive and negative facets were confirmed by our empirical findings. Furthermore, findings from multiple studies in our program of research consistently showed that the various technology beliefs can be categorized into four distinct components. Two of these components—*optimism* and *innovativeness*—are "contributors" that increase an individual's technology readiness, while the other two—*discomfort* and *insecurity*—are

EXHIBIT 3-4

DRIVERS OF TECHNOLOGY READINESS

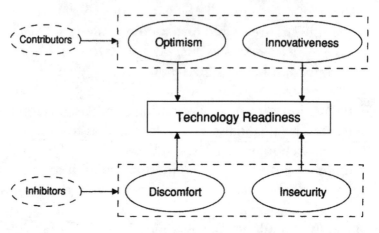

"inhibitors" that suppress technology readiness (see Exhibit 3–4).

Optimism

The optimism facet of TR can be defined as *a positive view of technology and a belief that it offers people increased control, flexibility, and efficiency in their lives.* It is a general dimension that captures specific feelings suggesting that "technology is a good thing." Ten of the statements in our 36-item TRI fall under this dimension. The following statements illustrate the types of beliefs contributing to optimism:

- You like the idea of doing business via computers because you are not limited to regular business hours.

- Technology gives people more control over their daily lives.

- Technology makes you more efficient in your occupation.

Exhibit 3–5 provides a pictorial profile of the distribution of the U.S. population on the optimism dimension. The skewed distribution, with a mean optimism score of 3.8 on the 5-point scale, implies that Americans are generally positive about technology. However, the long left tail of the distribution also suggests there is a segment (albeit a minority) of the population that is skeptical of the benefits of technology.

EXHIBIT 3-5

DISTRIBUTION OF THE NTRS SAMPLE ON OPTIMISM

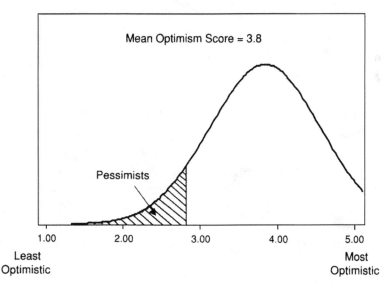

Technology optimists abound in the United States. Americans believe that technology allows them to accomplish more than in the past, and also gives them greater convenience and control. Specifically, findings from the NTRS show that

- 80% of the people surveyed believe that they are doing more with technology today than just a couple of years ago.

- 74% believe that technology gives them more freedom of mobility.

- 72% find new technologies to be mentally stimulating.

- 71% believe technology makes them more efficient on the job.

- 66% believe that products and services with the newest technologies are more convenient to use.

Technology optimism reflected by the aforementioned findings varies somewhat by age. Older adults, particularly those 65 or older, tend to be less positive about technology. For example, senior citizens (age 65+) are less likely to be trying new technologies or to find them to be mentally stimulating. Likewise, they find the notion of extending business hours by providing access through computers much less appealing than do younger adults.

NTRS findings pertaining to optimism also suggest that American consumers generally prefer capability over extreme simplicity in technology-based products and services. For instance, a vast majority of consumers (80%) prefer computer programs that they can tailor to their own needs, and about 60% prefer to have the most advanced technology in the products/services they use. Nevertheless, as implied by the long left tail of the distribution in Exhibit 3–5, our findings also suggest that enthusiasm for technology is by no means universal or unqualified. There is a healthy dose of pessimism about at least some facets of technology. For example:

- 93% of consumers want the benefits of technology demonstrated before buying.

- 90% prefer to talk to a person rather than a machine when they call a business.

- 83% believe that people often become too dependent on technology to do things for them.

- 47% find that technology designed to make life easier usually has disappointing results.

Older consumers are more prone to possess outright skepticism about technology. They are far more likely than others to believe the benefits of technology are grossly overstated, and a much greater proportion agree with the last statement, that technology may produce disappointing results.

Overall, considering both positive and negative views pertaining to technology optimism, younger consumers are generally more positive than older consumers on *some* facets of this dimension. Male and female consumers have similar views: both are quite positive about technology, but both also have some reservations, particularly in their belief that it is difficult to leave the human touch completely out of providing high-tech service.

Innovativeness

The innovativeness facet of TR refers to *a tendency to be a technology pioneer and thought leader.* It measures the extent to which an individual believes that he or she is at the forefront of trying out new technology-based products/services, and is considered by others as an opinion leader on technology-related issues. The 36-item TRI contains seven innovativeness statements such as the two shown below.

- You can usually figure out new high-tech products and services without help from others.

- In general, you are among the first in your circle of friends to acquire new technology when it appears.

Exhibit 3–6 profiles the population on the basis of their innovativeness scores. This distribution is fairly symmetric about the midpoint of the 5-point innovativeness scale, implying that the U.S. population as a whole has a modest level of innovativeness, with roughly one half above average and the other half below average on this overall dimension. Findings from the NTRS pertaining to various sub-facets of innovativeness reveal some interesting similarities and differences across the population.

EXHIBIT 3-6

DISTRIBUTION OF THE NTRS
SAMPLE ON INNOVATIVENESS

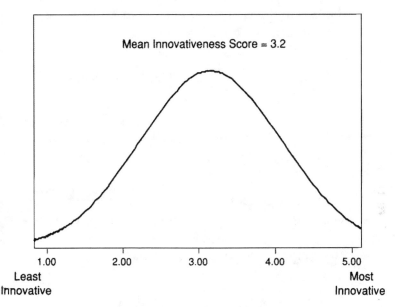

Mean Innovativeness Score = 3.2

| 1.00 | 2.00 | 3.00 | 4.00 | 5.00 |

Least   Most
Innovative   Innovative

For instance, our findings suggest that Americans are by and large receptive to new technologies.

- 87% believe that they are always open to learning about new and different technologies.

- 85% believe that learning about technology can be as rewarding as the technology itself.

These two views exist uniformly among both males and females, and among different age groups—the only exception is that a somewhat lower percentage of the 65 and over age group (72%) is open to learning about new technologies.

About two-thirds of the adults surveyed have the following innovative characteristics:

- 64% keep up with the latest technological developments in the areas that interest them.

- 63% enjoy the challenge of figuring out high-tech gadgets.

However, only about half the consumers over the age of 50 possess these traits. Moreover, males are somewhat more likely than females to keep up with the latest developments (69% vs. 60%) and enjoy the challenge of tackling new technologies (73% vs. 55%).

On the other hand, certain innovativeness-related beliefs are held by much smaller percentages of consumers. These characteristics define "true" innovators falling on the right half of the distribution in Exhibit 3–6. The following findings from the NTRS illustrate these characteristics:

- 50% think they can figure out new technology-based products and services on their own.

- 36% believe that others come to them for advice on new technologies.

- 32% say they are the first among their friends to acquire new technology when it appears.

- 51% are of the opinion that people will miss out on the benefits of technology when they delay a purchase for something better to come along.

These core innovator characteristics are strongly related to gender and age. Males are somewhat more likely to figure out new technologies on their own (60%), give advice to others (43%), and be the first to try new technologies (37%). These tendencies are much less pronounced among older people, especially those over 50 years of age. For example, 45% of those in their 30s are "thought leaders," offering advice to others, while only 13% of the 65+ age group do so.

Discomfort

The discomfort facet is an inhibitor of TR and pertains to *a perceived lack of control over technology and a feeling of being overwhelmed by it.* It represents the extent to which people have a general paranoia about technology-based products and services, believing that they tend to be exclusionary, rather than inclusive of all kinds of people. The following two items from the TR scale are examples of statements relating to discomfort (the 36-item TRI contains 10 such statements):

- Sometimes you think that technology systems are not designed for use by ordinary people.

- When you get technical support from a provider of a hi-tech product or service, you sometimes feel as if you're being taken advantage of by someone who knows more than you do.

EXHIBIT 3-7

DISTRIBUTION OF THE NTRS SAMPLE ON DISCOMFORT

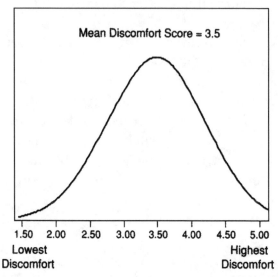

As Exhibit 3-7 shows, the distribution of scores reflecting the degree of discomfort Americans experience with technology is bell-shaped and centered slightly above the midpoint of the scale; people fall within the entire spectrum. A perceived lack of control over technology is a significant facet of discomfort, as illustrated by the following findings from the NTRS:

- 61% get overwhelmed by how much they need to know to use the latest technology.

- 45% believe technology is often too complicated to be useful.

These views are somewhat more widespread among

females and older consumers. For example, more females (51%) than males (38%) believe that technology is often too complicated to be useful; this belief is also much more prevalent among the 65+ age group (65%) than among the under-30 age group (30%).

Another important driver of discomfort is the perceived lack of effective technical support services. A vast majority (92%) of consumers consider it helpful to have a new high-tech product or service explained to them by a knowledgeable employee; in contrast, only about half of them find tech-support services to be adequate.

- 53% believe that tech-support personnel do not explain things in understandable terms.

- 47% believe that tech support sometimes makes them feel as if they are being taken advantage of.

- 53% think there is no such thing as a manual for a high-tech product or service that is written in plain language.

Similar to the views pertaining to the perceived lack of control over technology, the beliefs concerning tech-support services are somewhat more prevalent among females and older consumers.

The beliefs discussed thus far imply that a great deal of technology-related discomfort exists among American consumers. However, as suggested by the left half of the distribution in Exhibit 3–7, there is also a sizable segment of consumers whose discomfort with technology is not that high. The following findings illustrate the kinds of beliefs held by this segment:

- 50% of American consumers believe they have fewer problems than others in making technology work for them.

- 46% prefer to solve technology-related problems on their own rather than call for help.

- 45% claim they are usually in control of new technologies.

Mirroring the gender- and age-related differences noted previously, these views are more widely held by males and younger consumers.

Insecurity

Insecurity, another inhibitor of TR, can be defined as *distrust of technology and skepticism about its ability to work properly.* Though somewhat related to the discomfort dimension, this facet differs from it in that it focuses on specific aspects of technology-based transactions rather than on a lack of comfort with technology in general. Illustrative of insecurity items in the 36-item TRI (which contains nine such items) are the following statements:

- You do not consider it safe giving out a credit card number over a computer.

- You do not feel confident doing business with a place that can only be reached online.

- If you provide information to a machine or over the Internet, you can never be sure if it really gets to the right place.

Findings from the NTRS show that the American public as a whole experiences a great deal of technology-related anxiety. The distribution of insecurity scores (shown in Exhibit 3–8) is skewed toward a high degree of insecurity, with a mean value of 4.0 on the 5-point scale. As illustrated by the following findings, feelings of insecurity are especially widespread in the context of the Internet and e-commerce:

- 87% believe that any business transaction conducted electronically should be confirmed in writing.

EXHIBIT 3–8

DISTRIBUTION OF THE NTRS SAMPLE ON INSECURITY

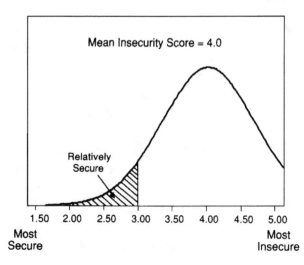

Mean Insecurity Score = 4.0

Relatively Secure

1.50 2.00 2.50 3.00 3.50 4.00 4.50 5.00

Most Secure

Most Insecure

- 70% worry that information they send over the Internet will be seen by other people.

Moreover, as noted in Chapter 2, insecurity is also reflected by a high degree of concern with giving out credit card numbers over computers and dealing with businesses accessible only online.

Although people's technology-related insecurities are most pronounced for online transactions, they are by no means limited to e-commerce. The public also has a certain degree of skepticism concerning technology in general and the ability of nonhuman processes to operate properly. For instance:

- 82% believe that when something gets automated, one must check carefully to see if the machine or computer is making mistakes.

- 78% believe that switching to a revolutionary technology too quickly can be risky.

- 62% say that technology always seems to fail at the worst possible time.

In contrast to the gender- and age-related differences observed for the other dimensions of technology readiness, the views pertaining to insecurity by and large vary little across gender and age groups. The apparently universal nature of these beliefs adds to the importance of insecurity as a serious inhibitor of people's technology readiness, except perhaps in the case of the small minority (represented by the narrow left tail of the distribution in Exhibit 3–8), who have few reservations and feel confident about technology's ability to deliver.

Distinctiveness of the TR Dimensions

The results show only modest associations among the facets. The four facets of TR are fairly independent, with each making a unique contribution to overall technology readiness. We conducted various statistical analyses to ascertain the extent of overlap among the four facets of technology readiness, and the results show only modest associations among them.[15] Moreover, even the modest associations we observed are limited to those between the *contributor* facets of optimism and innovativeness, and the *inhibitor* facets of discomfort and insecurity. The associations *across* the contributor and inhibitor categories (e.g., the association between optimism and insecurity) are much weaker.

The evidence of strong independence between the contributors and inhibitors of TR is consistent with the previously introduced notion that individuals may *simultaneously* harbor positive and negative beliefs about technology. Thus, high scorers on the contributor facets are not necessarily low scorers on the inhibitor facets. In other words, even technology optimists and innovators may apparently experience technology-related anxieties at levels similar to those experienced by individuals who are much less enthusiastic about technology to begin with. This lack of redundancy across facets has important implications for segmenting people on the basis of their technology readiness, a topic we discuss in subsequent chapters.

EXHIBIT 3-9

TECHNOLOGY READINESS INDEX: DISTRIBUTION

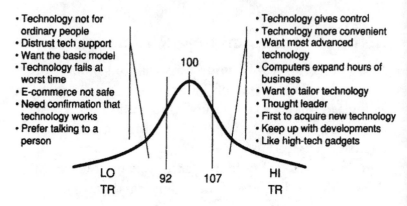

TECHNOLOGY READINESS INDEX

The combination of scores on the four dimensions represents a person's overall technology readiness. The shape of the distribution of overall TR scores is similar to that of a typical bell curve (representing a "normal" distribution with 50% of the scores below the mean and the remaining 50% above the mean). Exhibit 3-9 presents a standardized version of this distribution in which the mean TR score is scaled to an index of 100 for the sake of expositional convenience. The bell curve in this exhibit also shows some distinguishing traits revealed by the NTRS for the lower third of consumers (those with TRI scores less than 92) and for the upper third of consumers (those with TRI scores greater than 107).

The pattern of TRI scores across countries, and across different regions or markets within countries, may well vary from that shown in Exhibit 3-9. Nevertheless, the overall TRI distribution in Exhibit 3–9, as well as the distributions of TR-facet scores in Exhibits 3–5, 3–6, 3–7, and 3–8, can serve as benchmarks against which results from other market- or region-specific TR studies can be compared to yield potentially useful insights. The battery of 36 TR items is robust enough to be adapted for use in those studies. Researchers in other countries are also using the scale, and, over time, results from their studies will shed light on differences across cultures (e.g., is the Americans' seemingly passionate interest in the latest technology a cultural or a human trait?). In the next section we examine the soundness of the TR scale in terms of its ability to make meaningful distinctions among different levels of ownership and usage of technology-based products and services.

Link between TRI Scores and Technology-Related Behaviors

The NTRS, in addition to containing the battery of TRI statements, included a variety of other questions that probed respondents about their current and future technology-related behaviors. We examined the correspondence between respondents' TRI scores and their behavioral data obtained from three types of questions:

1. Owning or having in-home access to various technology-based products/services (e.g., cable TV, Internet service). People told us not only what they currently own, but what they plan to acquire in the near future.

2. Engaging in specific technology-related behaviors, usually in a service context (e.g., using an ATM, making online purchases). Here, people told us about their current behaviors and plans for the future.

3. Desirability of engaging in a variety of "futuristic" technology-based activities (e.g., making full two-way videophone calls, using robot checkouts at supermarkets). Respondents rated each service on a 6-point desirability scale (1 = *Very undesirable*, 6 = *Very desirable*).

The results from these analyses offer insight into issues such as the following:

• Is the TRI able to distinguish well between users and nonusers of high-technology services?

• Is the TRI's ability to distinguish between the two groups stronger for more complex or more futuristic technologies?

• Is the TRI's ability to distinguish between the two groups stronger for technology-based services for which user discomfort and insecurity are likely to be especially germane or pronounced?

If the TRI is indeed a sound measure of the technology readiness construct, answers to the questions above should be in the affirmative. Exhibit 3–10 contains bar charts representing mean TRI scores grouped according to three levels of ownership of (or subscriptions to) eight different technology-based products/services. Of

EXHIBIT 3-10

MEAN TR SCORES FOR DIFFERENT GROUPS OF OWNERS/SUBSCRIBERS

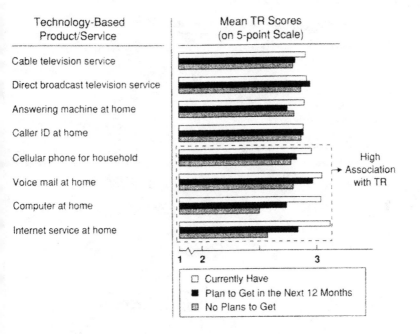

| Technology-Based Product/Service | Mean TR Scores (on 5-point Scale) |

- Cable television service
- Direct broadcast television service
- Answering machine at home
- Caller ID at home
- Cellular phone for household
- Voice mail at home
- Computer at home
- Internet service at home

High → Association with TR

1 2 3

☐ Currently Have
■ Plan to Get in the Next 12 Months
▨ No Plans to Get

these eight products/services, the first four are arguably easier to use and hence less complex from the users' perspective than are the remaining four. Moreover, they do not call for as much technological savvy and user involvement to operate properly; as such, user discomfort and insecurity are less likely to be critical issues for them. As evidenced by the bar charts in Exhibit 3–10, the mean TRI scores are similar across the three ownership/subscribership categories for the first four products/services: cable TV, direct broadcast TV, answering machine, caller ID.

EXHIBIT 3-11

MEAN TR SCORES FOR DIFFERENT GROUPS OF USERS OF TECHNOLOGY-BASED SERVICES

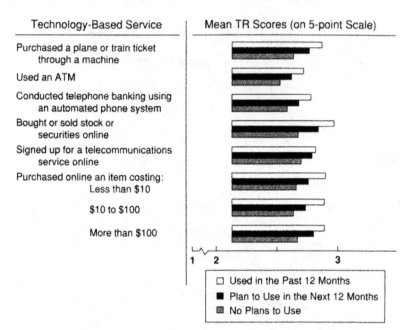

Technology-Based Service	Mean TR Scores (on 5-point Scale)
Purchased a plane or train ticket through a machine	
Used an ATM	
Conducted telephone banking using an automated phone system	
Bought or sold stock or securities online	
Signed up for a telecommunications service online	
Purchased online an item costing: Less than $10	
$10 to $100	
More than $100	

□ Used in the Past 12 Months
■ Plan to Use in the Next 12 Months
▨ No Plans to Use

In contrast, for the four remaining products/services—cellular phones, voice mail, computer, and Internet service—there is a *consistent* pattern of variation. Consumers already having the product/service have higher TRI scores than do those just planning to get it in the next 12 months; and, the latter group of consumers, in turn, have higher TRI scores than do those with no plans to get the product/service. The consistency of these findings with the a priori expectation that technology readiness is especially germane for products/services that are complex from the users' perspective lends credence to the TRI's ability to predict acquisitions of advanced technologies.

Exhibit 3-11 provides a pictorial summary of the correspondence between the TRI scores and *actual usage* (as opposed to just ownership/subscribership) of several technology-based services. It is reasonable to expect that customers with a low inherent propensity to embrace and use new technologies are likely to be intimidated by the kinds of services listed in Exhibit 3-11: purchasing a plane/train ticket through a machine, using an ATM, conducting banking transactions over an automated phone system, buying/selling stock/securities online, signing up for a telecommunications service online, and purchasing products online. Usage of these services is thus likely to be strongly associated with technology readiness. The bar chart patterns shown in Exhibit 3-11, without exception, are consistent with this expectation: for each service, current users are more technology ready than nonusers just planning to use it in the next 12 months, who, in turn, are more technology ready than those having no plans to use the service. These consistent patterns of results further point to the association between TR and technology behaviors, and the ability of the TRI scale to capture this relationship.

Technology readiness is also strongly associated with consumers' "technology vision," or how they would like to use technology in the future. The TRI scores correlate highly with people's *potential* technology-related behavior, measured as the extent to which they would find various high-tech products and services desirable in the future. Exhibit 3-12 charts results from our analyses pertaining to the desirability—as perceived by high-,

EXHIBIT 3-12

LINK BETWEEN TR AND DESIRABILITY OF TECHNOLOGY-BASED SERVICES

Technology-Based Service	Mean Desirability Scores (on 6-point Scale)
Maintain a family home page on the Internet	
Use a robot checkout at the supermarket	
Watch an interactive television that allows customization of program content	
Vote in a local-government referendum from a home computer	
Purchase small items like tickets to events over the Internet	
Purchase a large item like a car or furniture over the Internet	
Make phone call with full two-way video	
Send a voice message over the Internet	
Visit the World Wide Web through WebTV rather than a computer	
Attend an online class that allows electronic information exchange among all parties	
Read a book off a CD or the Internet with the aid of a portable electronic viewer	
Allow a computer to help diagnose and treat a medical problem	
Apply for a large loan over the Internet	
Own an emergency beacon for identifying a person's location	

1 2   3   4   5

■ High TR
■ Medium TR
□ Low TR

medium-, and low-TR consumers—of 14 different futuristic or recently introduced technology-based services. Again, almost without exception, perceived desirability scores for low-TR consumers are lower than that for medium-TR consumers, which, in turn, are lower than that for high-TR consumers. The fact that the perceived desirability of emerging technology-based services differs consistently—and in the predicted directions—across the three TR segments adds to one's confidence in the story the TRI scale reveals.

SUMMARY

Science requires tools to study real-world phenomena. An astronomer might rely on a telescope, a physicist on a particle accelerator, a geneticist on a map of the human genome. Social scientists frequently rely on scales to study human attributes and beliefs. The TRI scale serves as our tool for studying consumer behaviors for technology.

In this chapter we outlined the conceptual underpinnings of the technology-readiness construct, described a scale (the TRI) for measuring it, and discussed findings (from the National Technology Readiness Survey) pertaining to the scale's effectiveness as a tool for predicting and explaining technology-related behaviors. As demonstrated by the NTRS findings, people's propensity to embrace technology varies widely, resulting from an interplay between contributors (optimism, innovativeness) and inhibitors (discomfort, insecurity) of technology readiness. The TRI is an example of a reliable multiple-item scale that companies can use to gain an in-depth understanding of the technology readiness of their customers as well as their employees. The TRI scale can help address a number of issues, which are outlined below (subsequent chapters discuss these and other related subjects in greater depth).

By examining the TRI scores of current customers, a company can help answer a variety of questions pertinent to its technology strategies. For instance, what is our customer base's overall level of readiness to interact effectively with technology-based products and services? How does it compare with the technology readiness of the public at large (as measured by the NTRS)? Are there distinct segments in our customer base that differ in terms of technology readiness? If so, what are the relative sizes of those segments, and do they have any distinguishing demographic, lifestyle, or purchasing characteristics? Answers to these questions can provide useful insights pertaining to issues such as the extent to which technology-based systems should be the conduit for customer-company interactions, the types of systems that are likely to be most appropriate, the pace at which the systems should/could be implemented, and the types of support needed to assist customers experiencing problems with technology-based systems.

Researchers can also use tools like the TRI scale to assess the technology readiness of employees (i.e., internal customers). As in the case of external customers, gaining a good understanding of the technology readiness of employees is important for making the right choices when designing internal technology systems intended to help employees better serve external customers. The issue of technology readiness is especially important for contact employees to whom customers may turn for assistance when they encounter difficulties in dealing with a company's technology-based interfaces. In those situations, satisfactory problem resolution will hinge not only on the contact employees' people skills but also on their technology readiness. Employees who rate high on both interpersonal skills and technology readiness are likely to be much more effective in tech-support roles than are employees who are deficient on either criterion. As such, the TRI can serve as a supplementary screening device, along with traditional people-skills assessments, in selecting personnel for tech-support positions.

Later chapters will discuss strategies for marketing and servicing technology-based products and services. These strategies are based on a combination of insights from formal studies using the TRI as well as our observations as consultants to clients introducing technology to their markets.

CHAPTER 4

The Five Types of Technology Customers

When a new technology is introduced to the market, consumers will react in different ways depending on their beliefs. In the previous chapter, we identified four dimensions of beliefs that influence technology adoption. It is important to recognize that these four dimensions are *independent,* such that an individual can possess any combination of motivations or inhibitions.

The fact that an individual is driven to adopt technology in one area does not mean that he or she is equally driven in another area, nor does it mean that he or she would lack inhibitions. A person can be a technology "innovator," prone to experimentation, but be skeptical about the value of technology, or can believe strongly in technology but also fear it.

An output of our research is a *typology* of technology consumers. The market can be segmented into five distinct groups with different combinations of innovativeness, optimism, discomfort, and insecurity. The segments range in size from 14% to 27% of the adult population in the United States. They are distinct not only in their beliefs about technology but also in their perspectives on life and their demographic composition. Each segment plays a distinct role in moving a new technology-based product to maturity, entering the market at different times with different agendas.

The adoption of technology is analogous to the settling of a frontier. The first people to arrive are "Explorers," who are highly motivated and fearless. The next to arrive are "Pioneers," who desire the benefits of the new frontier but are more practical about the difficulties and dangers. The next wave consists of two groups: "Skeptics," who need to be convinced of the benefits of settling the frontier, and "Paranoids," who are convinced of the fruits but unusually concerned about the risks. The last group, "Laggards," may never move unless they are forced to do so. (See Exhibit 4–1.)

Exhibit 4–2 shows the technology beliefs of each segment in comparison to the general population. **Explorers** are extremely high in technology readiness, ranking higher on drivers and lower on inhibitors of adoption. They are an easy group to attract when a new technology is introduced; they will comprise the first wave of customers.

Laggards are the opposite of Explorers, ranking lower in motivation and higher in inhibition than the market as a whole. Laggards are the last group to adopt new technology. Indeed, they may do so only because they

EXHIBIT 4-1

TECNOLOGY ADOPTION SEGMENTS

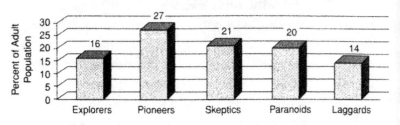

have no choice. For example, Laggards may find that vinyl record albums are no longer available in the store, forcing them to buy tapes or compact disks; or, they may find that their bank has cut back teller hours after introducing ATMs, leaving them no choice but to withdraw money from machines.

The middle three segments (in terms of their overall technology readiness) have more complicated beliefs about technology. **Pioneers** share the optimism and

EXHIBIT 4-2

DIFFERING BELIEFS OF
TECHNOLOGY ADOPTION SEGMENTS

	Drivers		Inhibitors	
	Optimism	**Innovativeness**	**Discomfort**	**Insecurity**
Explorers	High	High	Low	Low
Pioneers	High	High	High	High
Skeptics	Low	Low	Low	Low
Paranoids	High	Low	High	High
Laggards	Low	Low	High	High

innovative tendencies of the Explorers, but they have a certain degree of discomfort and insecurity. They are drawn to the use of technology but will encounter obstacles in the process that require the attention of the marketer. They need help making the technology work for them and require some degree of assurance. Pioneers are usually the next group in line to try technology.

Skeptics and Paranoids usually follow the Pioneers in acquiring a new technology. The reasons each waits longer are entirely different. **Skeptics** are dispassionate. They simply do not believe strongly in technology and lack any desire for pure innovation. They do not loathe technology; rather, their level of optimism is slightly below the market average. Further, Skeptics lack inhibition, ranking low in discomfort and insecurity. A marketer must convince Skeptics that the new product or service will benefit them. Once they believe, adoption can come readily because there are few reasons to hold back.

A **Paranoid** is like a child encountering a burning candle for the first time. The flame is fascinating, but also frightening and painful to touch. These consumers are optimistic about technology. They lack a tendency to innovate. More important, they exhibit a high degree of discomfort and insecurity. They need little convincing by the techno-ready marketer that they will benefit from the product. What they really need is support and reassurance.

Each segment enters a technology market at different phases. Exhibit 4–3 shows a timeline for home access to the Internet and online services. This service category continues to grow rapidly, with potential to reach a high penetration level in the United States,

EXHIBIT 4-3

TIME PERIOD WHEN TECHNOLOGY SEGMENTS REACHED 10 PERCENT PENETRATION OF INTERNET ACCESS

		Skeptics		
Explorers	Pioneers		Paranoids	Laggards
7/95	10/96	5/97	1/98	9/98

much like the telephone or the automobile. The time-line shows when each segment reached a penetration level of 10%,[1] giving a sense of market timing. Explorers reached a 10% level of penetration in July 1995, while each of the remaining groups passed this milestone over a period of about three years.

For an offering that is new and technology-intensive, the bulk of sales will come from *first-time users*. This is in contrast to a mature product or service category, where most of the sales are derived from people switching brands, replacing units, or increasing units in the household. Over time, the characteristics of the first-time users of a techno-product or service will change markedly, since their unique beliefs will determine their timing in becoming a buyer.

In contrast, the first-time users of a mature product that is no longer technology driven will tend to be people who are ready to consume because they have come of age for the product. Marketers of mature products and services are obsessed with life cycles. In today's market, first-time users include children for soft drinks, teens for automobiles, people in their 20s for real estate, and retirees for senior living communities. Technology readiness and its components do nothing to define the

EXHIBIT 4-4

NEW CUSTOMER COMPOSITION BY AGE OF TECHNOLOGY

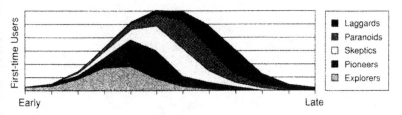

market for a mature product or service; technology beliefs mean everything for a techno-offering.

The technology segments described here provide a framework for describing who customers are at given points in the introduction of a technology. As illustrated in Exhibit 4-4, new customers in the beginning will be primarily Explorers. Eventually, the dominant segment will be Pioneers, then Skeptics and Paranoids. In the case of Internet and online services, the new customers have been predominantly Explorers up until 1997. Since 1998, the dominant segment has consisted of Pioneers. Other segments have been growing rapidly as the Pioneer segment becomes saturated and slows in growth. By 2001, Skeptics and Paranoids will probably dominate the market of newcomers for Internet/online services.

IMPLICATIONS OF THE
CUSTOMER TYPOLOGY

The distinct timing of different segments for a technology-based product or service has strategic implications for

1. During different stages of product introduction, the market focus must shift to reflect the needs and concerns of the people signing up at the time.

2. By proactively focusing on the themes of the emerging segments, a marketer can accelerate growth. The particular offering of the company that does this has potential to outpace competitive offerings since it appeals to the newest buyers.

3. Pioneers are one of the most critical segments. A new technology can be highly successful among the elite group of Explorers, but can fall on its face while still a niche product. The Pioneers, with their baggage of insecurity and discomfort, must be added to the customer list to propel a brand toward critical mass.

Later chapters provide a detailed prescription for marketing technology-based products and services, taking into account our emerging knowledge of consumer behavior. For now, a discussion of the broad strategies required over the life cycle of a new technology is in order. These strategies are illustrated in Exhibit 4–5, which depicts the dominant segments at different stages of market maturity, the prevailing "theme" in the market, and broad strategies for success. The following explores in greater depth the strategic implications for techno-driven marketing at each stage.

Early Adoption Stage

In the earliest stages of a market, the dominant segment is the Explorer, the dominant theme *innovation*. Mar-

EXHIBIT 4-5

THE SHIFTING FOCUS OF A TECHNOLOGY-BASED PRODUCT OR SERVICE

age of velopment:	Early Adoption	Accelerating Growth	Peak Growth		Declining Growth
minant gment:	Explorer	Pioneer	Skeptic	Paranoid	Laggard
rket emes:	Innovation	Discomfort and insecurity	Low optimism	Discomfort and insecurity	Market maturity and resistance by "hold-outs"
rategy:	Target Innovators, make products future-ready, and build a market base	Focus on usability and reassurance	Promote product benefits	Increase focus on usability and reassurance	Focus on retention and innovations

keters will deliberately target the most innovative con-sumers in order to achieve short-term sales. If they go further, they can nurture this early group of customers, turning them into "evangelists" for the company's tech-nology. This is akin to the strategy pursued by Apple Computer in its early years (discussed in Chapter 7). In anticipation of rapid change, this is also the best time to study the needs of techno-ready consumers to glean ideas for making the technology future-ready.

Accelerating Growth Stage

As market growth accelerates, the new customers will shift to that critical segment of Pioneers, who really need the product but are not as savvy as the first mar-ket entrants. At this stage, the product or service must be *customer-focused*. This means building ease of use into product design. Leading companies will subject their products to rigorous usability testing. Customer support becomes a key focus (e.g., customers with discomfort issues are more responsive to help from a human voice). At this stage of increased growth, it is important to let the user know that the product is safe and to design technology-based products and services that are just that. It is not enough that it work properly; reassuring the user that everything is okay is also important.

Managers in charge of technology-based products and services in this stage of marketing could be deceived about the nature of their most recent customers, whom they may refer to as "newbies." There is often more to these customers than merely being first-time users of a technology. Newbies seem to have special issues—problems using the service, worry about its safety, a cautious show-me attitude concerning what it can do for them. These attitudes are not necessarily due to the level of experience, but to the stage of adoption. Indeed, if the marketer thinks back to a couple of years earlier, he or she may recall that those first-time customers were remarkably self-sufficient, confident, and firm in their convictions. The marketer may also be surprised that the newbies of today will not change much in the future. If these newbies have problems with the product now, they will have problems with it later unless it is modified.

In one of our consulting assignments for a new customer-service Web site, we presented our client with wonderful news. The first wave of customers had given the site extremely high marks, particularly in critical areas such as "navigational ease" and "look and feel."

Clearly, much of this was a result of an unusually thoughtful and well-planned development process.

Our news, however, was tentative. We measured the technology readiness of these early customers and discovered that they were in the 78th percentile among the online population (indeed, they were even higher compared to the general adult population that includes people not online). These new customers were clearly Explorers. The application was new for the company, and these innovative customers had rushed to sign up to do all of their business with them online. The client had feared that these customers would be extremely critical. Instead, they mirrored what we have found in research on service quality of technology-based services—techno-ready consumers tend to be more satisfied, not less.

We forewarned the client that the next wave of customers may not have as easy a time using the site, and would need more help. This was confirmed by an interesting operational statistic. As the number of customers signing up to use the site climbed, so did the percentage who contacted technical support with problems. The implication was that the telephone technical support would play a critical role in establishing a connection with the newest customers.

Peak Growth Stage

When a technology is at its peak growth (when the rate of growth is at its highest before slowing), different themes will dominate. At this point, many new prospects will need to be convinced of the merits of a new technology. By accident, these Skeptics may realize the benefits on their own. For example, a key reason in the past for buying a wireless phone was the desire to have one on hand for road emergencies. Countless numbers of first-time consumers getting service for this reason soon discovered other benefits of the technology. It gave them newfound freedom and accessibility.

A company can capture Skeptics by emphasizing the particular benefits of its own brand. High-speed Internet access can free time for other family members to go online. A network for a small business can save money on software and phone lines. A machine that dispenses travel tickets saves a wait in line. However, it helps to communicate these benefits to the market. Some companies are known for their ability to present benefits of emerging technologies in vivid detail, and many exciting demonstrations can be seen at the International Consumer Electronics Show in Las Vegas. For example, in 2000, Sun Microsystems and GTE provided a live demonstration of home network technology at their GTE Connected Family™ Home. They displayed a Web cam of a home in Dallas, and showed how they could conduct tasks such as dimming the lights on command, driving home the tangible benefit of peace of mind when traveling.

As the market matures, discomfort and insecurity will continue to be important themes. Consumers may begin to trust the safety and reliability of a new technology, since it will have been around for awhile. But the newest consumers will be even more concerned than the "newbies" of the past. More of these new customers will be Paranoids whose perceptual barriers of discomfort and insecurity will be the most formidable yet encountered. Apple Computer responded to this theme when it introduced its iMac, a leapfrog in design simplicity that appealed to first-time buyers in a market that had been around for over two decades.[2]

Declining Growth Stage

In the final stages of a technology, growth will continue, but at a declining rate. At this stage, marketers can pursue different options. Potentially, a company could target Laggards, the consumers who are least inclined toward any new technology. Yet new opportunities arise with any technology-based product or service as a result of technological advancements. These advancements will change the nature of the game, starting the cycle anew to attract Explorers and Pioneers, who are in their perpetual quest for new advantages. Even an established technology like the television has promise for a new revolution. Digital technology provides opportunities for more vivid sound and images, interactive capabilities, and expanded channels, spurring a whole new market cycle.

Another focus in this later stage should be on building customer loyalty and focusing on re-sales and secondary sales. Gateway provides an excellent example of a model for turning a product into a relationship. The company created a program called "Your Ware," which allowed consumers to obtain a customized package of bundled hardware, software, and Internet solutions and to pay a flat monthly fee. The program also included obsolescence protection, allowing customers to trade in their hardware after two years.[3]

Much of the growth in maturing technologies comes from the sale of additional units within households. A large share of computer sales are now either replacements or units for other family members. Family members can now call each other when away from home on their own portable phones.

Summing it up, companies can ride the product curve, responding to the unique needs of each of the five segments in the typology of technology consumers. These segments have thus far been defined by their beliefs about technology, but they are also distinctive on other characteristics. The next chapter provides a more in-depth profile of each segment, giving us a better understanding of their behavior.

CHAPTER 5

A Closer Look at Technology Customers

Each technology segment discussed in Chapter 4 has a distinct personality and background. The segments themselves are determined by a pattern of beliefs about technology, but there are also differences in demographics, values, and product usage. Even within a segment, consumers are heterogeneous. A segment may tend to include younger consumers but could claim a certain portion of senior citizens. Its members could tend to be female, but they could also include some males. Even so, the differences are significant enough from segment to segment to help a marketer gain more familiarity with the kinds of consumers who dominate each stage of a product's life cycle. These are summarized in Exhibit 5–1 and discussed in detail in the paragraphs that follow.

EXHIBIT 5-1

PROFILE OF TECHNOLOGY SEGMENTS

Segment	Explorer	Pioneer	Skeptic	Paranoid	Laggard
Technology Readiness Index (100=Average)	123	104	102	88	79
Technology Beliefs	True believer and innovative	Motivated but hampered	Needs to be convinced	Insecure	Low motivat and high barriers
Demographics	Young, upscale, male	Young, middle class	Average age, upper middle class	Middle age, lower income	Older, lower income
Psychographics	Curious about world	Impulsive, success-oriented	Deliberate	Under pressure, self-conscious	Lacks curios
Technology	Heavy user	Heavy user	Moderate user	Light user	Light user

Explorers, usually the first to acquire an innovative technology, are unique among all the segments. An Explorer is defined as being highly optimistic about technology, innovative, comfortable in its mastery, and secure in its relative safety. Nearly all American adults enjoy technology to some degree, but Explorers are particularly fond of trying new toys. They find technology mentally stimulating. Indeed, they seem prone to obtain cutting-edge products as much for the fun they provide as for their utilitarian value. At the same time, Explorers believe in the ability of technology to provide them freedom, control, and efficiency. (See Exihibit 5-2.)

Explorers are classic early adopters, the first in their circle of friends to try new things. They also stay informed of cutting-edge developments in technology.

EXHIBIT 5-2

EXPLORERS

16% of U.S. Adults
TR Index: 123 (100 is average)

Technology Belief Patterns

Most optimistic
Most innovative
Least uncomfortable
Least insecure

Representative Opinions about Technology*

You like computer programs that allow you to tailor things to fit your own needs (98% agreement)
You find you have fewer problems than other people in making technology work for you (81% agreement)
You keep up with the latest technological developments in your areas of interest (91% agreement)
You find new technologies to be mentally stimulating (93%)

Demographics

Highest education (40% have college degrees)
Highest income (44% have annual household incomes of at least $50,000)
Youngest (Median age 36 years, 58% under age 40)
More likely to be a student
Most likely to be male (62%)
Most likely to be in a technology profession (45%)

Psychographics

Like to explore world and try new things
Prefer cerebral pursuits to physical activity
Success-oriented

Technology-Based Products and Services (>50% Ownership)

Computers, cell phones, caller ID, ATMs, online services, telephone banking

Agreement defined as "strongly" or "somewhat" agree.

They are thought leaders and perceive that others seek them out for advice. As a fountain of information, they have the potential to influence the acceptance of a new product.

Explorers are confident in their ability to make technology work for them. If anyone can program a VCR, it is an Explorer. Explorers would prefer to take the time to customize their own software so it meets their precise needs. This is a mark of self-confidence, since other consumers may tend to avoid tinkering and accept the default options chosen by the manufacturer. Explorers need less help from technical support and are less likely to feel they have problems with technology. They trust new technologies more than other consumers and perceive less risk to their security.

Almost half of Explorers work in a technology profession. Occupation is either a sign that they are interested in technology, a contributor to their interest, or perhaps both. These adults have the highest incomes of any segment and are the most educated (40% have college degrees, compared to less than a fourth of the general population). Explorers are the youngest of any segment, though "young" is relative. While many are under age 30, half are age 36 or older. Two-thirds are male (though a smart marketer does not overlook the fact that many are female).

True to their name, Explorers are curious about the world around them and like to learn about new things. They are more likely than others to be thinkers, preferring cerebral activity to physical pursuits. These consumers are also concerned with having a successful career. Explorers are avid consumers of technology-based products and services. Most own computers, have Internet access at work and home, and even without computers will use automated services such as telephone banking. They embrace communications services such as cell phones and caller ID, and will bypass human interaction in favor of ATMs (now being enhanced by online banking).

Summing it up, technology is both a means and an end to Explorers. They lead lives that involve technology in life and work, satisfying their desires for learning and play. But technology also plays a role in their success, since they gravitate to highly paid technology professions.

Pioneers are above average in technology readiness and are usually the next in line after Explorers to adopt technology. Comprising a fourth of the market, this group possesses a high level of optimism about technology and is somewhat innovative. Yet Pioneers are more practical and discerning and will be less forgiving of technology that thwarts them or fails to reassure. (See Exhibit 5-3.)

Pioneers recognize the benefits of technology and want the latest and greatest. They usually desire the most advanced options available to them, and they believe that technology makes them efficient in their occupations. Like Explorers, they derive enjoyment from their technology, find it to be mentally stimulating, and enjoy learning about it. Pioneers are the only segment besides Explorers who happen to be innovators. These consumers are the first to try new things. They keep up with the latest developments and enjoy gadgets. Half consider themselves to be thought leaders.

Adopting technology is somewhat painful for Pioneers. They definitely have more issues than Explorers and have slightly more resistance to technology than the market as a whole. They tend to believe that technology systems are not designed for "ordinary people." They perceive a certain degree of risk in paying too much for

EXHIBIT 5-3

PIONEERS

27% of U.S. Adults
TR Index: 104 (100 is average)

Technology Belief Patterns
More optimistic
More innovative
Greater discomfort
More insecure

Representative Opinions about Technology*
Technology makes you more efficient in your occupation (81% agreement)
Learning about technology can be as rewarding as the technology itself
(90% agreement)
If you provide information to a machine or over the Internet, you can never be sure it
really gets to the right place (80% agreement)
Whenever something gets automated, you need to check carefully that the machine
or computer is not making mistakes (88% agreement)

Demographics
Average education (21% have college degrees)
Average income (25% have annual household incomes of at least $50,000)
Relatively young (median age 39 years, 51% under age 40)
Half are male (54%)
More likely to be in a technology profession (24%)

Psychographics
Like to try new things, including toys and gadgets
Impulsive
Success-oriented

Technology-Based Products and Services (>50% Ownership)
Computers, cell phones, caller ID, ATMs, online services

Agreement defined as "strongly" or "somewhat" agree.

technology that does not prove its worth. Sharing a characteristic of other consumers with discomfort about technology, they possess a degree of distrust of technical support.

Pioneers need some reassurance about new technology. They are average in their general fears of technology. For example, half believe that technology dangers are often overstated, the same as the general public. Pioneers are more concerned than typical consumers about new information technologies. They fear that information they send over the Internet will be seen by other people, and they would be concerned giving out a credit card number over the computer.

Pioneers do not stand out much from others in terms of demographics. They are about average in income and education, and they split evenly between male and female. They tend to be younger than other segments (except for Explorers), with half being under age 40. They are more likely than others to work in technology professions, and they have a higher incidence of computer access in their jobs.

These consumers seem to derive pure joy from technology, with a personality trait of favoring toys and gadgets. They admit to being impulsive and trying things on the spur of the moment. Like the Explorers, Pioneers consider a successful career to be an important life goal. Pioneers have a high incidence of ownership of technology-based products and services (second only to Explorers). More than half possess computers, cellular phones, caller ID, and online services, and most use ATMs.

Pioneers are important consumers to a marketer of technology because of their positive views and willingness to innovate. Technology is a key to their life goal of success, but they are not as self-reliant as the first wave of adopters, and they will require more reassurance about technology.

Skeptics have few inhibitions to adopting technology. They are relatively comfortable in using it and secure

EXHIBIT 5-4

SKEPTICS

21% of U.S. Adults
TR Index: 102 (100 is average)

Technology Belief Patterns

Less optimistic

Less innovative

Less discomfort

Less insecure

Representative Opinions about Technology*

If you provide information to a machine or over the Internet, you can never be sure it really gets to the right place (20% agreement)

Other people come to you for advice on new technologies (5% agreement)

Demographics

Average education (22% have college degrees)

Average income (33% have annual household incomes of at least $50,000)

Average age (median age 40 years)

Half are male (52%)

Less likely to be in a technology profession (12%)

Psychographics

Not impulsive

Less likely to take control of life

Technology-Based Products and Services (>50% Ownership)

Computers, ATMs

Agreement defined as "strongly" or "somewhat" agree.

that it works safely and reliably, but these consumers wait longer to make an acquisition. They have only a moderate degree of optimism and virtually no desire to be innovators. With this mindset, they will probably wait until the benefits of a new techno-offering are proven and demonstrated to them. (See Exhibit 5-4.)

Skeptics are not against technology, only less enthusiastic than those who enter the market sooner. For example, only half believe technology gives them more control over their lives, while this belief is widespread among Explorers and Pioneers. Skeptics believe that technology is convenient to use and makes them efficient, but they do not feel as strongly as the segments who lead the market. Skeptics lack any tendency toward innovation. They rarely find themselves in the role of giving others advice. They usually believe that their friends adopt new technology before they do. In effect, these consumers sit back and wait while others they know prove the value of new technologies for them.

Skeptics have few problems mastering technology and rarely believe it is too complex. While half of all adults sometimes feel that they are taken advantage of by technical support, few Skeptics perceive such a problem. Skeptics are also more comfortable with manuals for technology-based products. Most, however, get overwhelmed with the challenge of staying on top of the latest technologies, and less than half feel they have fewer problems with technology than others.

These consumers are skeptical about the benefits of technology but not about security. They perceive far less risk than typical technology users. For example, they are less likely than any segment except Explorers to believe that technology will fail them at the worst time. They are unconcerned with the security risks of online commerce, and they rarely doubt that information sent through a machine or the Internet will fail to reach its destination.

Demographically, Skeptics are average, middle-class people. They mirror the general population in education and age, though they have a somewhat higher income than the average consumer. Half are male, the majority are married, and almost half have children at home. They are less likely than others to be in a profession related to technology. As might be expected from individuals who lack innovator traits, Skeptics tend to be less impulsive than others. They are more prone to sit back and let fate take its course rather than leap into something new.

Unlike Explorers and Pioneers, Skeptics lack a blind faith in technology. They will let others blaze a path before they acquire a new technology, and only then because they are convinced of its value.

Paranoids believe in technology but are held back from adoption by strong inhibitions and, to a lesser extent, a lack of innovative tendencies. As a result, they usually enter a technology market at a relatively late stage, when growth begins to slow. Their overall technology readiness is below that of Skeptics, which makes them more likely to adopt later. In some situations, however, they may keep apace of this other group, and marketers would need to contend with both segments at about the same time. (See Exhibit 5–5.)

Paranoids share the vision with Pioneers and Explorers that advanced technology can improve their circumstances. Two-thirds believe that technology gives them more control over their lives. They believe technology makes them more efficient workers and is more convenient to use. At the same time, they are devoid of any tendency to innovate solely for the sake of innovating. They rarely dispense technology advice, have trouble figuring out technology without help, and find their friends trying things before they do.

What really slows Paranoids from trying advanced technology is their high level of discomfort and their

EXHIBIT 5-5

PARANOIDS

20% of U.S. Adults
TR Index: 88 (100 is average)

Technology Belief Patterns

More optimistic
Less innovative
Greatest discomfort
Most insecure

Representative Opinions about Technology*

It seems your friends are learning more about the newest technologies than you are (82% agreement)
The human touch is very important when doing business with a company (92% agreement)
It is embarrassing when you have trouble with a high-tech gadget while people are watching (88% agreement)

Demographics

Less educated (12% have college degrees)
Below average income (16% have annual household incomes of at least $50,000)
Older (median age 45 years, 41% at least age 50)
Mostly female (63%)

Psychographics

Feel pressure from fast pace of life
More likely to take control of life
Conscious of own image and symbolism of products they buy
Brand loyal

Technology-Based Products and Services (>50% Ownership)

ATMs

Agreement defined as "strongly" or "somewhat" agree.

especially high level of insecurity. More than anybody, they believe that technology will fail them at the worst possible time. They are concerned with replacing important people-tasks with automation because of the potential for error. These consumers believe that many new technologies pose health or safety risks that will only be discovered after they are in use.

Almost every consumer, no matter how techno-ready, tends to believe that the "human touch" is important, but Paranoids are the most passionate in this belief. They find technology overwhelming, complicated to use, and wonder if technology systems are really designed for use by ordinary people.

Paranoids might be described as the "struggling middle class." Compared to the population at large, they are slightly older, more likely to be female, and less affluent. They are usually middle-age—half are 45 years or older—and two-thirds are female. Paranoids have lower incomes and lower educational levels than other segments. They are the least likely to own a business and less prone than others to work in technology professions.

These consumers feel pressure in life, wishing its pace would slow. Their cautious view about technology is reinforced by a tendency to want to take control of their affairs rather than leaving things to chance. Thus, while the hesitation of some segments may be a result of a passive approach of letting others act first, Paranoids may be proactively delaying until they are certain there are no risks. These individuals are self-conscious; they care what others think about them, and they believe that the products they use convey a message to others about themselves. It is quite possible that one of the risks of technology failure is social, the fear of appearing foolish when a product does not work for them.

Paranoids are among the lowest-incidence users of technology-based products and services. They are barely getting used to ATMs and have not yet achieved a high penetration of items like cellular phones or computers. Although they believe in the value of technology, they deliberately hesitate. Their themes are insecurity and discomfort. As a techno-driven product or service approaches maturity, a marketer must sell Skeptics on the value of technology, while reassuring the Paranoids.

Laggards are the last people to adopt a new technology, and everything about them works against doing so. This includes a lack of faith in benefits, insecurity, discomfort, and a habit of letting others adopt first. They are also inhibited by a low disposable income and a lifestyle free from pressure to stay ahead. (See Exhibit 5–6.)

Their lack of belief in advanced technology is striking. Unlike other consumers, they doubt that technology can deliver control, convenience, or efficiency. The majority of consumers in every other segment appreciate the potential of computers to expand the hours they can do business, but not the Laggards; they are often retired and can operate on "banker's hours." Like the Skeptics and Paranoids, Laggards have virtually no tendency to be natural innovators. They let others acquire knowledge, disseminate knowledge, and experiment. The majority of consumers like high-tech toys and gadgets, but not the Laggards.

To the extent Laggards can comprehend technology-based products and services, they question whether they will operate securely and are wary of sending information over the Internet. They also distrust technical support. If they purchase a high-tech product, they will opt for the basic model in order to keep things simple. Laggards are conscious of the potential dangers of revo-

EXHIBIT 5-6

LAGGARDS

14% of U.S. Adults
TR Index: 79 (100 is average)

Technology Belief Patterns

Less optimistic

Less innovative

Greatest discomfort

Most insecure

Representative Opinions about Technology*

If you buy a high-tech product or service, you prefer to have the basic model over one with a lot of extra features (69% agreement)

You do not consider it safe to do any kind of financial business online (80% agreement)

In general, you are among the first in your circle of friends to acquire new technology when it appears (15% agreement)

Demographics

Less educated (17% have college degrees)

Below average income (18% have annual household incomes of at least $50,000)

Older (median age 56 years, 57% at least age 50)

Mostly female (67%)

Most are retired or not working full-time (58%)

Psychographics

Not prone to keep up with fashion

Not success oriented

Brand loyal

Doesn't like to try new things

Technology-Based Products and Services (>50% Ownership)

Not a heavy user of technology-based products and services

Agreement defined as "strongly" or "somewhat" agree.

lutionary technology and are more likely to believe the negative press that warns of these dangers.

A Laggard is likely to be "Grandma" (though many are "Grandpas" as well). Laggards are the oldest of any segment, with the majority being over age 50, and two-thirds are female. Many are retired, and less than half work full-time. Age does not guarantee a lack of interest in technology—a fifth of Explorers are over age 50—but it is certainly an important correlate. Perhaps adults lose their desire to do things differently as they age, or maybe retirement removes pressures younger adults experience. Retirees do not need to worry as much about working and shopping efficiently. Only a third of Laggards have children at home, so there is less need to use technology such as computers and the Internet for education. They earn less money than typical house-holds, reducing their ability to invest in technology. They are also less educated.

Laggards lack a desire to explore and learn new things. They also show little penchant for high-tech toys and gadgets. They are not particularly interested in keep-ing up with new styles and fashions. These consumers are the hardest sell for a technology marketer, and it begs the question: Why even try? In a fast-changing technolo-gy market, the best bet is to move on to new opportuni-ties once the potential among Skeptics and Paranoids is tapped out.

In concluding this discussion on technology seg-ments, it is important to stress that there is no such thing as the perfect paradigm for describing a marketplace. To illustrate, experts may argue over the validity of supply-side versus Keynsian economics, but a smart deci-sion maker would regard these theories as models, not truths, and draw from both to predict the economy. The same holds for a theory of technology consumer behav-ior. Our typology is based on empirical research. It goes beyond other theories by adding new dimensions and more data. There are, however, other ways of viewing a technology market. Furthermore, our research is ongoing, our model evolving, so what we describe today will undoubtedly change in the future. Our typology is nonetheless a useful way of looking at the world, so we can now turn to the next section of this book, which discusses implications for marketing techno-driven products and services.

CHAPTER 6

The Pyramid
Model of Marketing

Effective marketing of technology-based innovations requires a thorough understanding of not only the typology of TR-based customer segments—the focus of Chapters 4 and 5—but also the technology-triggered changes in the ways companies interact with and serve customers—the focus of this chapter. The advent of new technologies, most notably the World Wide Web, is drastically altering the modus operandi of marketing organizations. Marketers in many sectors, including those wherein technology traditionally has not been all that important (e.g., hotels, catalogs, magazine publishers, education), are finding new avenues for communicating with customers, cultivating customer loyalty, distributing their wares, and offering customer service.

Historically, the focus of the marketing function has primarily been on seller-buyer exchanges pertaining to physical products, from canned goods and cars in consumer markets to chemicals and construction machinery in business-to-business markets. As such, traditional marketing theories and concepts—captured by the so-called 4 Ps (product, promotion, price, and place, or physical distribution)—have been dominated by a goods orientation. Several significant trends over the last two decades, however, suggest that this conventional, goods-oriented form of marketing is inadequate for achieving *sustainable* competitive advantage:

- The proliferation of look-alike competitive offerings in virtually all product sectors

- The almost instantaneous imitation by competitors of a company's price-based promotions

- The increasing demand from customers for superior service

These trends signal a strong and clear message, one that savvy marketers are already heeding: enduring marketing success stems from *serving* customers well, not just *selling* to them. Providing any service—be it a core service such as offering investment advice, a peripheral service such as answering queries about a product shipment, or a supplementary service such as after-sales maintenance for a machine—invariably involves interactions between customers and company personnel. Recognizing the human-intensive nature of service provision and the opportunities it offers for fostering customer loyalty, marketing experts have suggested broadening the scope of conventional marketing to encompass the role of customer-contact employees.[1] Specifically, these experts urge companies to consider three types of marketing:

- **Internal Marketing:** This deals with "marketing" to employees, that is, making them strong believers in the company and motivating them to serve customers well. It requires treating employees as *internal* customers and providing them the support systems (e.g., training, tools, empowerment, incentives) they need to provide superior service to external customers.

- **Interactive Marketing:** This type of marketing pertains to fostering customer loyalty during employee-customer interactions. It calls for all customer-contact employees, regardless of their functional affiliation, to recognize and capitalize on opportunities to make a favorable impression on customers.

- **External Marketing:** This term captures a company's efforts to design, price, promote, and distribute its offerings. It represents the traditional 4 Ps of conventional marketing.

Exhibit 6–1 presents a triangle model of marketing that pictorially represents the three facets of marketing. *External marketing*—the only kind most companies think of as "marketing"—primarily involves making promises to customers and attracting them to a company's offerings. However, delivering on the promises and retaining customers are difficult to accomplish in the

EXHIBIT 6–1

TRIANGLE MODEL OF MARKETING

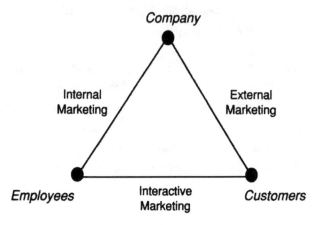

absence of effective internal and interactive marketing. As such, the triangle model underscores the fact that the role that marketing—even when low or no technology is involved—is broader and more complex than what conventional thinking might suggest.

The expanding role of technology in selling to and serving customers introduces yet another layer of marketing complexity. To capture this added complexity we propose the pyramid model of marketing shown in Exhibit 6–2.[2]

The introduction of technology as a service-delivery option adds a whole new dimension to marketing, mirroring the shift from the two-dimensional triangle model to the three-dimensional pyramid model. As we discuss next, technology has profound implications and offers exciting opportunities for managing the company-employee, employee-customer, and company-customer links.

EXHIBIT 6-2

PYRAMID MODEL OF MARKETING

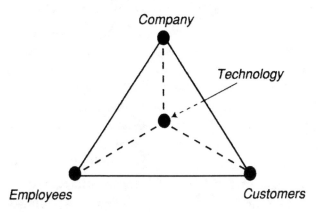

THE COMPANY-EMPLOYEE LINK

A silent revolution is taking place in how technology is being leveraged inside the marketing organization to facilitate intracompany communication and *internal marketing*. Some of this is very obvious, including the widespread use of information technologies such as e-mail, fax, and teleconferencing. According to the NTRS, 44% of the adult public had Internet access at work in 1999. Indeed, technology is driving some fundamental changes inside corporations:

- In a research study on corporate meetings we discovered that the boom in technology is increasing the demand for large, in-person meetings. The factor driving these meetings is the increasing tendency to distribute work across locations around the country and around the globe. Meetings bring this scattered workforce together face to face.

- Another study among CFOs revealed that the "annual budget" is becoming a relic of the past. More and more companies are using real-time budgeting software to continuously fine-tune resources to adapt to changes in their markets.

Technology facilitates marketing success from the inside of the organization in a number of ways. Newfound capabilities for more efficient communication serve to align employees and management, ensuring that they strive toward a common vision. Technology helps to coordinate diverse functions and departments whose collaboration is key to efficient marketing and optimal synergy. This newfound ability to coordinate increases in importance as organizations themselves become more complex. Technology also provides employees with timely information about customers and markets.

- More and more firms are building internal networks and Intranets, offering opportunities to link together diverse sources of information to create marketing synergy. Companies now have an opportunity to manage information dissemination, ensuring a consistent and motivating message to all employees. Departments can create their own home pages, and a company or division can have its own Intranet "start page" that is updated and tailored as needed, telling employees about new products, marketing campaigns, service initiatives, and the like. One of our clients in the entertainment industry used their Intranet to manage marketing activities in an extremely complex organizational structure. Creating a new television program, movie, or CD-ROM title had ramifications far beyond production. They had to inform a range of internal audiences for purposes of public relations, advertising sales, distribution, licensing and cross-channel promotions. The Information Technology (IT) department developed and tested an automated calendar that was used by each functional area to identify major activities and milestones, replacing the need for constant memos, phone calls, and e-mails.

- Using technology as an internal marketing communications tool is not restricted to large companies with fat budgets. Intranets.com is a free Web site that allows organizations to create their own communications Web site for use by internal communities and even collaboration with external audiences. Registered users have access to resources to conduct online discussions, store contact information, share documents, maintain a group calendar, and send updates to team members. Examples of some applications are linking up a geographically dispersed sales force, or coordinating "virtual" companies consisting of a loose network of independent businesses.

- Technology can be used to keep management up-to-date on their state of customer satisfaction. One of our clients is a leading provider of contract services with over 1,000 accounts around the country. The organization is deeply committed to quality and conducts an extensive client satisfaction survey. Survey results are posted onto a proprietary web site accessible to any of its approximately 150 top managers involved in client relations. Technology speeds up the dissemination of critical information. For example, an account with a high level of dissatisfaction is immediately flagged to four different management levels.

THE EMPLOYEE-CUSTOMER LINK

Interactive marketing refers to the traditional dialogue between employee and customer in the service relationship. This is an important connection, since consumers are by nature focused on human relationships. On the one hand, technology displaces some of the activity traditionally performed by staff. A machine or Web site can take over functions traditionally performed by ticket agents, bank tellers, telephone reps, or checkout clerks. On the other hand, technology offers some promising new avenues for employees in how they communicate and add value.

Many firms now offer the ability to communicate by an electronic medium such as e-mail or a dedicated online link. This is a norm for complex business relationships or technology services such as Internet access. For example, an AOL customer can get answers to questions by e-mail or engage in a conversation with an employee via online chat. Technology is becoming more important in businesses that are relationship focused. For example, the authors surveyed real estate agents on behalf of Interealty, a company that provides technology to multiple-listing services. The study found that 64% of real estate professionals use the Internet for business. Among those with Internet access, 44% have a personal

Web site and 20% communicate daily with buyers and sellers.[3]

Technology is also finding its way into the sales interaction. For instance, mortgage loan officers can complete an application on a computer in a prospective borrower's home, obtain a credit report online, and get the loan approved. In turn, the mortgage banker can connect electronically with Freddie Mac, a mortgage financier, and obtain a commitment to sell the loan with assistance from an automated credit scoring system.

As customers demand more interaction through a technology medium, the technology readiness of employees becomes an issue. The NTRS reveals that 48% of U.S. workers find it desirable to interact with customers by e-mail. Not every employee is ready to plunge head first into the online world or use computers in front of their customers. Human resource professionals will need to take into account the unique issues surrounding technology beliefs in decisions for recruiting, training, and job design.

An interesting outcome of technology's role in the service relationship is greater freedom for employees to customize service delivery and find creative solutions to customers' problems. Instead of handling routine transactions, customer relations staff add value by offering judgment, initiative, and charm. For example, a financial consultant at a brokerage firm is under greater pressure to offer true counsel, since the order taking can be handled by the Internet. The most successful employees are those who can leverage technology available to them to solve problems. To illustrate, a savvy customer service representative, equipped with a sophisticated database, can find the root cause of recurring problems or analyze a customer's purchase history to find a superior spending plan.

THE COMPANY-CUSTOMER LINK

Technology provides more conduits for company-customer interactions (*external marketing*), offering added convenience, flexibility, and lower cost. And technology facilitates different types of interactions. First, new information technology provides new ways to *communicate* with customers. Not too long ago, communication occurred primarily through mass advertising, employee interactions, or first-class mail. Today, every major corporation, and increasingly smaller businesses and nonprofits, are introducing expansive Web sites that create a more intensive interaction, one more capable of building brand equity and educating the customer.

- Sallie Mae, a leading provider of student loan financing, uses its Web site to educate students and parents about financial planning and debt management. It includes content and tools for parents, students, and the educational community. Sallie Mae is a secondary market that works behind the scenes while banks process its loans, making the World Wide Web one important tool in building brand equity.

- The Discovery Channel, a television network offering thought-provoking programming on science and nature, allows its Web site to be an entertainment product synergistic with its other businesses. Visitors can access original content for research and fun, check programming schedules, or interact with other fans.

- The Web site for Tide detergent is designed to encourage repeat visits and strengthen customer relationships. Features include links to family fashion news, tips and timesavers for laundry, contests, and tools cast in an entertaining format such as the "Stain Detective" to help consumers figure out how to remove difficult stains. It provides an unprecedented medium for a goods manufacturer, Procter & Gamble, to build relationships with consumers who previously interacted only with the retailer.

Another function of technology in the company-customer link is *customer service*. Companies have been leveraging technology for years to service relationships. Computerized telephone systems are not a new phenomenon, although a boom is occurring in Web-based servicing. For example, the NTRS shows that one out of five consumers have checked bank account information online, and one out of 10 have checked information for a utility account. Over time, the reliance on customer service representatives and counter employees has been decreasing as customers service their own needs through computers. The benefits include a wider time window, freedom from travel, speed, and postage savings.

- MCI WorldCom, one of the nation's largest long-distance carriers, allows customers to check their long-distance bill, pay their bill online, or sign up for service at their web site.

- Kaiser Permanente, a leading HMO, allows members to refill prescriptions through an automated telephone line.

- Marriott allows its guests to search for hotel locations, make room reservations, and get directions and maps to its hotels online. Guests who belong to its Rewards program can check their accounts and redeem travel points through the site.

Perhaps the most profound role of technology is as a *real-time channel of delivery*. By definition, this occurs with services that vend information or offer an information-intensive service such as banking. An application that has been around for years is the automated teller machine. Increasingly, the Web is being used to deliver information traditionally provided by a nonelectronic medium.

- Charles Schwab, a leading brokerage firm, was an early entrant into the area of online trading. Even more traditional firms, such as Merrill Lynch, have put aside concerns such as the role of the broker and plunged into this promising new area.

- Verizon, a telecommunications giant and publisher of Yellow Pages directories, allows consumers to look up businesses anywhere in the country. Consumers can throw away their heavy phone books and go to Big Yellow to get phone numbers, read ads, access maps, and shop.

- According to the NTRS, a fourth of consumers have or plan to sign up for an online publication. The NTRS also predicts that a major growth business will be online education (about half of the public find this to be a desirable technology activity).

The foregoing discussion highlights the tremendous potential that technology holds for enhancing the efficiency and effectiveness of the company-employee, employee-customer, and company-customer relationships represented by the three sides of the triangle model. Realizing that potential, however, without falling prey to technology's pitfalls requires a deeper understanding of the pyramid model's implications for the three kinds of marketing.

The pyramid model, by explicitly incorporating technology as a service-delivery option, in effect calls for an expansion of the domains of the three types of marketing. For instance, traditional internal marketing—the one-dimensional line linking company and employee—is now depicted in the pyramid model as a two-dimensional triangle that interlinks company, employee, *and technology*. This triangle is one of the four facets of the pyramid model and represents "expanded internal marketing" (Exhibit 6–3).

EXPANDED INTERNAL MARKETING

The expanded internal marketing perspective shown in Exhibit 6–3 raises several issues pertaining to the management of the employee-technology and company-technology links. The employee-technology link in this

EXHIBIT 6-3

EXPANDED INTERNAL MARKETING

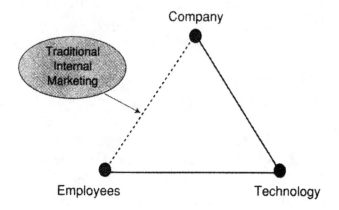

perspective symbolizes the need for companies to give careful thought to the following issues:

- Are systems in place to motivate our customer-contact employees to use the proposed technology effectively in serving customers?

- Do our customer-contact employees have adequate training in the proposed technology?

- Will they have ready access to the technology to take prompt corrective action when something goes awry?

Here is an illustrative scenario based on actual experience. A customer orders movie A electronically through his cable TV service and eagerly settles back to enjoy it. However, his TV begins receiving movie B instead. He immediately calls the cable company's customer-service department for assistance. The representative answering the phone, though polite and apologetic, is unable to rectify the problem because, according to him, there is no technician on duty and he himself has no idea about what to do except to cancel the cost of the movie, which has been automatically charged to the customer's account. In an ideal situation, the employee would have access to the same technology through which the movie was ordered and could correct the problem on the spot.

The company-technology link in the expanded internal-marketing perspective also raises important issues for management to consider:

- Is the proposed customer-contact technology appropriate for our target market?

- Are our target customers technologically ready to interact effectively with the technology?

- Are a sufficient number of properly trained personnel available to serve customers who may need assistance with the technology?

- Has the technology been evaluated by novices (as opposed to technically adept individuals)?

Hasty, haphazard introduction of customer-contact technologies can backfire by frustrating customers and fostering distrust. A case in point is in-home banking, an electronic, telephone-based service through which customers can conveniently complete conventional bank-

EXHIBIT 6-4

EXPANDED INTERACTIVE MARKETING

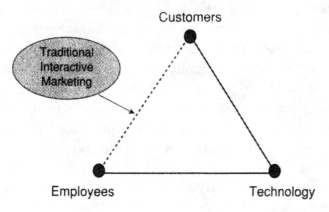

ing transactions by themselves. Despite being available in the United States for well over a decade, this hi-tech service is only now starting to reach a critical mass of market acceptance, in part due to low customer confidence and trust in it.

EXPANDED INTERACTIVE MARKETING

Expanded interactive marketing (Exhibit 6–4), another facet of the pyramid model, also has important implications for companies. Apart from the key issues already outlined (i.e., choosing the appropriate technology and providing employees with adequate access to and training in the technology), this facet involves addressing additional questions pertaining to the customer-technology link. The following questions illustrate:

- Are clear and friendly instructions available for customers who wish to use the technology?

Citibank, which pioneered automated teller machines over two decades ago, is constantly improving the customer-interface aspects of its ATMs through ongoing research with hundreds of customers who test and critique new models. The bank has now earned a reputation for having the most friendly ATMs. For instance, Citibank's ATMs in New York offer a choice of over a dozen languages in which customers can conduct their transactions.[4]

- In what ways can we "empower" the technology to customize the service delivery?

By creatively coupling artificial-intelligence systems with customer information, companies can give their customer-contact technology some degree of flexibility in serving customers (e.g., acknowledging long-term customers and providing them with special treatment to further strengthen their loyalty). The customer-service center of Bell Mobility Cellular, a major player in the North American cell-phone market, has "empowered" its automated call-routing system to recognize loyal, high-volume customers and move their calls to the front of the queue when the demand for service exceeds the center's capacity.[5]

EXPANDED EXTERNAL MARKETING

The implications of expanded external marketing (Exhibit 6–5), represented by the remaining two facets of the pyramid model, can be summed up in a single principal message for management: traditional company-to-customer communications (i.e., advertisements,

EXHIBIT 6-5

EXPANDED EXTERNAL MARKETING

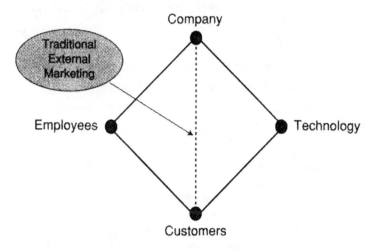

brochures, commercials, etc.) must now be especially sensitive to the capabilities—*and limitations*—of customer-contact technologies and employees. Specific questions that management should ponder in this regard include the following examples:

- Do our communications accurately inform customers of what our technology can and cannot do? Is there a danger that our communications may be promising too much (perhaps implicitly)?

- Do our communications articulate what customers should do when the technology malfunctions?

- Do our customers know whether and how they can get immediate hi-touch (i.e., employee) assistance if the hi-tech service fails?

TECHNOLOGY READINESS IN PERSPECTIVE

Exploring the linkages in the pyramid model highlights the need for giving careful thought to the technology readiness of customers and employees. Companies must leverage technology in a manner that is consistent with their own customers' level of TR. Thus, while companies must be receptive to working with cutting-edge technology to attract and retain customers, it is equally important for them to recognize and avoid potential technology pitfalls. As we discuss in the next two chapters, they must ensure that their strategies for marketing to and serving their customers explicitly take into account the technology readiness of their customers and employees.

CHAPTER 7

Acquiring
Technology Customers

I n our first chapter, we presented four principles for techno-ready marketing. The first of these principles holds that the model of consumer behavior for a techno-driven offering is different from that of a traditional product or service. Chapters 2 through 5 discussed the components of such a model based on our research.

The second principle of techno-ready marketing holds that commercial success depends on redefining marketing practices to reflect a revised model of consumer behavior. This chapter provides guidance on implementing the second principle, proposing specific practices that marketers of technology-based products and services should follow. Many of the practices we present have been used by leading companies, whose followers devised them based on their own consumer theory, intuition, or just trial and error. Our discussion provides case studies for companies that succeeded by adhering to these practices, and some that failed by ignoring them.

By following this second principle of techno-ready marketing, as well as the third principle—meeting the challenges of satisfying and servicing customers (the subject of the next chapter)—a company can rapidly accelerate the adoption of its offering while reducing costs of support. More important, marketers protect their long-term niche by being the first to achieve critical mass, the fourth principle of techno-ready marketing.

The following discussion presents four strategies for marketing a techno-based product or service:

- Technology evangelism

- Future-ready design

- Proving benefits

- Market-stage pricing

TECHNOLOGY EVANGELISM

The word *evangelism* connotes spreading a religious message. Perhaps this reference is too extreme; advanced technology can offer tangible benefits, but can it be compared to religion? At Apple Computer, evangelism is specifically used to refer to the spreading of a "platform religion." Apple, which practically created the personal computer industry, has had a stormy history, with ups and downs. It made many controversial decisions that are considered mistakes on hindsight; examples include keeping its operating system proprietary, antagonizing corporate IT professionals with its famous "lemmings" ad, and introducing product flops like the temperamental Newton MessagePad.[1]

What has sustained Apple through the decades is a near-religious faith in its technology. This zeal is not purely accidental; it has been propagated through a deliberate strategy of evangelism.[2] This strategy mirrors how a religious denomination might spread its doctrine, with the difference that Apple employs paid, full-time evangelists. The job of an evangelist is to champion a technology to users and developers. Like a missionary converting the masses, this sophisticated marketing professional will deliver an impassioned message to the media and public forums. Some of the evangelists, like Guy Kawasaki, achieved global recognition.

In its darkest days in the late 1990s, when profits plummeted and lay-offs soared, founder Steve Jobs returned to Apple as an evangelist. His role was to revitalize the company, introducing a new spirit of innovation, raising employee morale, and inspiring customers. At this time, the company relied even more on its cadre of missionaries to tell the world the company was making a comeback. They spread the word that Apple was not yet licked; it had a superior technology, a loyal customer base, and a healthy sales volume.

What ultimately makes evangelism successful, whether it pertains to a religion or a technology, is that one person converts another, who in turn converts

EXHIBIT 7-1

HOW TECHNOLOGY EVANGELISM WORKS

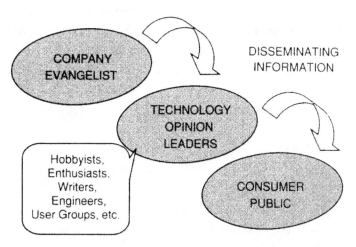

another, until the word is spread to a significant segment of the population (see Exhibit 7-1). Apple's evangelists, with all their fervor and determination, are merely a catalyst. The real work is performed by the loyal followers—customers, writers, distributors, engineers—who carry the banner for the company solely out of personal conviction, even when it is not their full-time job to crusade for this company's products. For example, in response to the negative publicity the company faced in its last crisis, a band of employees created their own Web site to tout good news about the company. Today, people can gather positive information from Web sites dedicated to Apple.[3]

We call technology evangelism an important techno-ready marketing strategy because it makes sense from a consumer behavior standpoint. It taps into one of the four dimensions of technology readiness, *innovativeness*. One aspect of this trait is a tendency to try new things. People with innovative behavior like to explore their world and play with toys, making them receptive to new technologies. Another key aspect of innovator behavior is a tendency to gather and disseminate information. Innovative people are interested in learning about new developments and telling others what they have learned. Whether they read about a fascinating technology in a magazine, or acquire the technology for themselves, they derive satisfaction from sharing their knowledge.

We have observed in countless research studies that the bulk of consumers who are *not* highly innovative have little fondness for gathering and assimilating information about technology. To illustrate, we commonly hear in focus groups about technology newbies relying solely on tech-savvy friends or relatives to help them make a purchase decision. An "expert" saves considerable time and pain by providing the newbies with a qualified solution. Innovative consumers, who are primarily the Explorers and Pioneers in our typology in Chapter 4, play an essential societal role of dispensing advice to consumers.

Apple is not the only company to practice technology evangelism; another master of this strategy is Sun Microsystems.[4] And, technology evangelism can take many forms besides a formalized corporate function. It can describe any strategy that harnesses the word-of-mouth power of consumers high in innovative tendencies.

- The growth in agricultural productivity in the United States can be attributed to the Agricultural Extension

Services whose agents sought out innovative farmers to adopt new techniques that would eventually be accepted more widely in the community.[5] Farming has been the subject of classic sociological research on the diffusion process for innovation.[6]

- A marketer may target a prestigious buyer in order to influence the marketplace. For example, a client of ours providing a systems product for colleges and universities sought out the leading institutions because it would make other campuses more receptive to their brand. In the browser wars between Netscape and Explorer, the two companies battled for well-known corporate clients for the same reasons.

- Software companies may release rough beta versions of their products to selected customers, even though the developer and customers know of flaws that need to be worked out. Certain users take pride in being the first, and they consider it a worthwhile trade for helping to debug the software.

- Companies may create forums of their own to disseminate propaganda, often providing free information that is appealing to consumers who like to stay abreast of trends. Common avenues include Web sites, paper newsletters, and e-mail newsletters. User groups play a role beyond gathering feedback for product improvements; they can be used as a way of energizing thought leaders. Best Software, a subsidiary of Sage Group, PLC, markets business application software for accounting, finance, and human resource functions. Its target customers are hard-to-reach, busy professionals, but Best fosters relationships by providing free information to benefit these professionals, including a national seminar series.

Just as a company can benefit from leveraging innovative consumers, it can also meet disaster by failing to do so. This was the fate of Divx, a home video system introduced by Digital Video Express. Divx was created through a partnership between Circuit City, a national electronics retailer, and a Los Angeles law firm. One of the main value propositions Divx offered was to save consumers the return trip to the video store. A consumer could purchase the disk for a small fee, about the same as a video rental, and watch movies as often as desired without having to return the disk. The system periodically updated viewing information over a phone line, billing the viewer's credit card for repeat usage.

Divx depended on market muscle to bolster its prospects. Its introduction was accompanied by a huge advertising budget, and it certainly did not hurt to be distributed through one of the country's largest retail chains. The format was initially made incompatible with DVD, or digital video disks, which were being introduced at the same time as Divx. This forced consumers to choose one over the other, creating fears of a format war. As a way of forcing consumers to opt for Divx, deals were struck with major movie studios to give Divx exclusive rights for blockbuster movies.

Ultimately, resistance in the marketing channel hammered Divx. Video stores were reluctant to carry its disks because it would cut into repeat visits, movie studios stopped supporting the medium, and only a few manufacturers would produce hardware. But the company's essential failure was how it antagonized the innovative consumers who were the first to acquire digital technology. Inflamed by the ill-conceived practices that limited choice, threatened uniformity, and appeared to violate privacy, innovators reacted with a vengeance. These thought leaders initiated a vehement campaign to discredit Divx. At one point, the Internet abounded with anti-Divx references. Divx might have avoided disaster if it had tested its concept with innovative consumers early on and adapted its product and message to win over thought leaders.[7]

Technology evangelism can be an effective strategy, particularly in the early market stages of a new technology when customers are Explorers and Pioneers with a propensity to innovate and proselytize. Evangelism is a natural dynamic that drives the early success or failure of any techno-driven product or service. If leveraged in a deliberate fashion, it will ensure a higher likelihood of success and faster growth.

FUTURE-READY DESIGN

One of the greatest risks a technology marketer faces is improperly timing the marketplace. Poor timing can include introducing products or services that are not sufficiently advanced for consumer tastes. Or, the offering can be too advanced, consisting of features that are over the heads of most buyers, possibly at the risk of driving up costs or complicating the design.

Why is timing so important when technology is

involved? Over time, cutting-edge features gain greater acceptance as consumers with lower levels of technology readiness decide to adopt them. Consequently, technology-driven markets are in a continual state of evolution as regards acceptance of more advanced functionality. For example, online shoppers may initially visit a Web site to gather information, but may be afraid to buy online. But in a matter of time, the same shoppers may develop a sufficient level of confidence to conduct purchases on the site. Over even more time, the shoppers may accept more advanced features: inputting a personal profile, adding themselves to a mailing list to hear about special bargains, or even delegating control to the commerce provider to send new supplies, refills, upgrades, and other things.

Exhibit 7–2 illustrates the role of timing. Technology offerings should be designed to be "future-ready." This means offering functionality that is in demand at the time of introduction, as well as including enough features to satisfy maturing consumer tastes until the next iteration is available. Making products future-ready poses a challenge for designers, since they are faced with a development cycle. Since technology markets change rapidly, a design that meets needs at the time it is on the drawing board could be obsolete by the time it reaches the public.

The dynamic nature of technology markets results in a different development process than for traditional markets. Instead of researching the market, identifying needs, and developing products to meet those needs, managers must use market research and judgment to predict what consumers will desire at the time of roll-out. In effect, marketers must attempt to predict the future.

EXHIBIT 7-2

FUTURE-READY DESIGN

gment	Explorers	Pioneers	Skeptics	Paranoids	Laggards
vel of hnology diness	High ←			→	Low
at they sire in hnology now	Cutting-edge functionality		Functionality that is in widespread use today		Obsolete technology
at they sire in hnology in future	Next-generation functionality (not available in the recent past)		What Explorers and Pioneers desired in the recent past		Obsolete technology

*ure-ready design is achieved by identifying product needs from the more technology-
dy segments of the market, Explorers and Pioneers. They will tend to desire cutting-edge
ctionality, which is more likely to be in widespread use in the near future. Skeptics and
anoids will tend to desire functionality that is currently in widespread use; relying too
ch on the advice of these less techno-ready segments could result in technology that
mes obsolete too quickly, while following the advice of the more techno-ready segments
more likely ensure that technology meets the needs of the general market in the near
ure.*

In our role as researchers and consultants, we have
often found ourselves in the position of evaluating a new
product or service that has an opportunity to be more
future-ready. Examples include the following:

- An entertainment Web site offered a Web cam that
featured a static picture of a location related to the
site topic. The picture was captured in real time and
could be refreshed to capture changes in action by
the click of a mouse. Users were disappointed. They
expected the picture to continually refresh itself, or
even offer live images, an increasingly common feature on websites.

- Introductions of blockbuster CD-ROM titles, such as Doom and Myst, changed expectations for the effects when a user moved through the virtual space created by the designer. Titles that offered a less advanced engine suffered in appeal, even if the content was interesting, since game players had come to expect a more lifelike experience.

- A new Internet service provider, faced with the need to compete with America Online, included an "instant messaging" feature allowing users to engage in real-time chats with others online at the same time. In testing, novice users were unfamiliar with and did not comprehend the value of such a feature, but more experienced users often looked for it without being asked. The provider recognized that the feature would come into increasing demand over time and made it a priority for the service.

- A cable company rushed to market a cable-modem technology that offered high-speed access only one way, for downloading information. Information that was uploaded, or sent by the user's computer, had to be sent over a standard telephone line at slower speeds. In focus groups, the distinction was lost on less experienced users but was a sore point for some of the more techno-ready users. Other cable companies have chosen to wait until they have upgraded their infrastructure to offer a more advanced system with two-way, high-speed access, anticipating that consumers would eventually develop more intensive usage patterns that would create a need for fast uploading as well as downloading.

History also offers interesting cases of failures that result from introducing technology that is not sufficiently advanced for the market. In the 1960s, many firms attempted to break into the IBM-dominated computer mainframe market, often failing because of inexperienced marketing and inability to offer advanced-enough technology. For example, Xerox entered the market by acquiring Scientific Data Systems in 1969, but faltered by the early 1970s as its products became obsolete.[8]

Another interesting example comes from the field of economic development. New York City at one time invested in the construction of new docks, but avoided the infrastructure for container shipping, a cost-saving innovation. One of the reasons cited for this ill-fated decision was political pressure from labor interests concerned about losing jobs. Shippers ended up relying on container facilities in New Jersey, causing the new docks to be abandoned.[9] In this case, the decision to avoid future-ready design was deliberate, and the results were disastrous for the city and the workers who depended on the jobs.

One could argue that technology can never be too advanced, and that the winner is always the company with the most prolific engineers and most advanced features. But such an argument ignores practical constraints faced by product designers. First, a richer array of features, particularly advanced ones, may result in added costs that the market is not willing to bear. Furthermore, incorporating more features may increase the development time, resulting in a wider window for competitive threats to emerge. There is also a balance that must be struck between the feasibility of an emerging technology feature and its demand in the marketplace.

More time may be required to work out bugs and add refinements, or the bulk of systems in use may not be fully compatible with the new technology.

Marketers are often presented with a rich new technology but are faced with tough decisions about exactly how to channel it into meeting consumer needs. A good example is the set of decisions faced by the television industry with the advent of broadband digital cable systems. Cable operators and other industry players have a plethora of choices as capacity to homes is expanded: increasing the number of cable channels, allowing consumers to choose programming whenever they desire, creating interactive programming, increasing information content, adding search capabilities, offering higher quality pictures and sound on new high-definition televisions, and even adding functions unrelated to television, such as home management. Decisions have to be made on how to allocate the new bandwidth, and the wrong decisions could result in costly, misplaced investments.

Failing to be future-ready is not the only mistake a company might make in terms of properly timing the introduction of technology features. A company may also stumble by offering a design that is too sophisticated. Buyers may not share the marketer's vision for what the product can accomplish. Worse, the sophistication may be a cause of the product or service's failure, driving up price, making the product complex, or causing confusion in the market.

Lotus is a case study in aspiring to a higher level of functionality than the market is ready to accept. The company's Lotus 123 software product was a "killer application" in the 1980s. The PC market was in its infancy, and the electronic spreadsheet was a revolutionary product that provided obvious value to millions of users. Lotus saw far into the future and recognized the potential for integrated applications, combining spreadsheets with word processing and database management programs to create a synergistic package with a seamless interface. This kind of integration is a given to computer users of today, who purchase software applications in suites and benefit from the integrating abilities of powerful operating systems like Windows. But in the early 1980s, their Symphony product proved to be too far ahead of its time and beyond the grasp of the typical user.[10]

Another example is Polavision, an instant movie camera system introduced by Polaroid in the late 1970s. The company foresaw the potential for a product that allowed consumers to record and play back moving pictures instantly, a function identical to the modern home video system. At a time when video recording was primarily an industrial technology, Polaroid introduced a system that processed home movies instantly. Though a cutting-edge technology, the Polaroid product was expensive, complicated to use, and of limited quality compared to the conventional 8-millimeter home movies of the time, which had to be sent out for processing. Unlike home video systems that came out a few years later, it was not possible to record over the movies, and the projection system did not have secondary uses such as playing rented videos. The product never gained the popularity of Polaroid's instant cameras and had to be discontinued after taking a $68.5 million write-off. Polavision was a far-thinking idea that was too far ahead of its time.[11]

Implementing a Future-Ready Design Strategy

The strategy of incorporating future readiness into a design seems an insurmountable challenge, since it requires the ability to predict the future. Fortunately, technology adoption follows a natural progression from the highest to lowest technology-ready consumer. It is impossible to predict even the near-term future without flaw, but a company can gain a better sense of the direction it should take by engaging in a dialogue with the more techno-ready consumer. Opinions should be sought from Explorers, who are first to take the plunge into any innovation, and with Pioneers, who are also early adopters but have a more practical perspective.

A disciplined marketing organization—in a technology-driven field or otherwise—will have a formal market research process for obtaining consumer feedback. The difference in approach is that the research for the company with an innovation should focus on the more techno-ready consumers. The process will explore the desired applications and current ways of using a particular techno-driven product or service. The research will ferret out two clues that imply a feature will rise in importance as the category matures: (1) the feature is in relatively widespread or growing use among the techno-ready segment of the market; and (2) interest in the feature correlates with techno-readiness, that is, there is a substantial gap in interest between the high- and low-techno-ready consumer.

The divergence of opinion and behavior across segments with different levels of techno-readiness is an important clue. It suggests that an application or feature is viewed through the filter of technology beliefs that distinguish a techno-offering from a more traditional good. Since the feature has acceptance among the techno-ready buyers, it stands a good chance of gaining acceptance in the future among other consumers. Why is this divergence a clue of future behavior in the market as a whole? More-techno-ready consumers engage in innovative behaviors sooner because they are more optimistic, more prone to experiment, and less resistant. Less-techno-ready consumers may engage in the same behaviors, but will wait longer because of their skepticism and apprehension. They may also wait until their more-techno-ready friends have tested technology and indicated their approval. The relationship between technology readiness and behavior applies not only to an entire innovation but also to the manner in which an innovation is used, making high-TR consumer behavior a good proxy for product and service design. As noted in an earlier example, many consumers may visit e-commerce sites initially, but the more-techno-ready will be the first to enter their credit card numbers online and buy.

It should be stressed that there are no guarantees when it comes to trying to predict future behavior. The product or feature that is in higher use among more techno-ready consumers may simply be an anomaly that never gains wider popularity. Or, a new technology may reach the market that supplants the future-ready feature.

Despite the pitfalls of predicting the future, technology readiness serves as a useful tool for identifying the demands of tomorrow. The National Technology Readiness Survey illustrates the future readiness of different technologies and technology features. Exhibit 7–3 plots a list of existing products and services by two dimensions, mirroring the criteria described above: (1) the level of usage among highly techno-ready consumers; and (2) the **Divergence Index,** which is the ratio of usage among high-techno-ready consumers to that among low-techno-ready consumers. To illustrate, the reported incidence of use of a subscription to an online publication is 26% among high-TR consumers and 13% among low-TR consumers; thus, the divergence index is 26 ÷ 13 = 2.0, indicating a relatively high level of association with technology readiness (a total absence of association would produce an index of 1.0).

The results tell a lot about the future potential of an innovation and whether its growth will follow a natural diffusion to lower techno-ready buyers. For example, ATMs are in widespread use, but usage is high among low- and high-techno-ready consumers alike. As part of a banking service package, ATM access is a strong feature but would not make a package future-ready. In contrast, online banking has a high future potential, as evidenced by the high Divergence Index. In addition, usage is already substantial among the highly techno-ready population—a third have engaged in this activity—indicating the application is not obscure or too cutting-edge to be practical.

In an actual product design process, consumers would be asked to rate their usage of or interest in specific features on a list. The designer could then examine and compare each of the features on the techno-readiness criterion. As a core requirement, the designer would want to include features that have widespread perceived value. In addition, the designer would use the

EXHIBIT 7-3

THE FUTURE-READY POTENTIAL OF DIFFERENT
TECHNOLOGY-DRIVEN PRODUCTS AND SERVICES

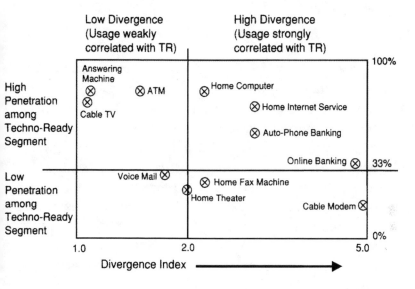

techno-readiness information to identify the special set of features with emerging potential that might make the product or service future-ready. These would tend to be features like those found in the upper-right-hand quadrant in Exhibit 7-3.

The key point is that the technology marketer is faced with the dilemma of having to predict the future in order to ensure that the final product is ready for the market at the time of introduction. One approach for determining what will be mainstream in the near future is to focus on highly technology-ready consumers. The customers the marketer should talk to include not only the early-adopting Explorers, but the

Pioneers, who possess a more practical, balanced view of technology.

PROVING BENEFITS

As a technology market approaches maturity, growth can be stalled by a failure among consumers to understand the benefits of the product or service. Consumers who have not yet acquired the technology may consider it an abstraction, a trendy idea that seems to be popular but for which there is no clear reason for adopting. In short, a large segment of consumers will fail to "get it" unless marketers take steps to prove the benefits of the technology, demonstrating how it can have a positive impact on the consumer's life. Despite this need to articulate benefits, the providers of advanced technologies sometimes fall short. They may be solid engineers versed in the technical advantages of their product, but they are unable to explain what it means in practical terms.

In the early market stages, proving benefits is less of an issue, since the purchasers are primarily Explorers and Pioneers. These consumers are inherently optimistic about technology and are able to connect its applications with their own lifestyle and work situation. Further, they are naturally innovative, meaning they are willing to experiment even if they are not totally certain of the benefits they will reap from their investment.

After Explorers and Pioneers, Skeptics become an important segment to ensure continued growth, particularly since they have the financial means for paying for new technology. These consumers will not take it on faith that the technology is a worthwhile investment of time and money, so it becomes critical for a marketer to prove to them that they, too, should adopt the techno-driven product or service.

High-speed Internet access serves as a good example of this need to prove benefits to consumers. With the introduction of high-speed lines to homes and business-es through technologies such as cable modems and DSL lines, the surfing experience is changing immeasurably. Because these technologies cost more, a consumer would not bother to use them unless they offered a clear benefit over a regular phone line. In talking to con-sumers in focus groups and surveys about the technolo-gy, we have found that highly techno-ready consumers immediately grasp the advantages of higher speed and are eager to acquire it if they have not already done so. In contrast, less-techno-ready consumers think that faster speed does little more than alleviate some frustra-tion when surfing, an inconvenience that is not worth the added cost.

A deeper exploration of the subject with consumers reveals concrete benefits for high-speed access. An example of such a benefit is family harmony. Long download times in using the Internet can drag out a ses-sion. Many families often have multiple users who fight for their turn to go online. By getting people off the computer faster, there is more peace in the household, a benefit that is easily understood.

Sometimes, the benefits are not understood until after a consumer begins to use a technology. Mobile telephones experienced rapid growth in the 1990s as costs dropped, units became more portable, and service improved. Many consumers acquired their phones for one reason: to be able to call for help in an emergency, such as the car breaking down on the road. They acquired the technology with the intent of using it infrequently. Soon, many of these consumers began placing and receiving calls for nonemergency reasons. They discovered that their phones offered greater freedom of mobility, accessibility, and control. For these consumers, the cell phone was transformed from a passive tool for emergencies to a necessity of life. Yet this success story could also indicate a lost opportunity. If marketers of mobile phone services had more aggressively communicated the profound impact that their service would have on peoples' lives, sales might have increased more rapidly a decade ago.

In the early 1900s, Thomas Edison used a demonstration approach called the Tone Test to prove the benefits of the phonograph to the public.[12] His company arranged for the Metropolitan Opera in New York to place a phonograph on stage with a live artist. The audience would then be challenged to distinguish whether they were listening to the recording or the real person. Memorex is famous for a similar demonstration with its "Is it live or is it Memorex?" ads, amazing national TV audiences with a Memorex tape recording of the singer Ella Fitzgerald shattering a wine goblet.[13] Similarly, marketers of digital television and DVD players are today finding ways to portray the vividness of their technology over conventional television, including ads designed to clarify the differences in image quality. Circuit City's Web site provides an in-depth consumer guide in plain language about the benefits of DVD for its customers; it describes features such as higher resolution, surround sound, and wide-screen views, using illustrations to demonstrate, for example, the difference between 320 and 720 pixels ("picture quality is twice as good").[14]

A company that excels at communicating the benefits of its technology to consumers is Select Comfort Corporation, a manufacturer and retailer of air beds. The company offers an innovative product line based on air-inflated mattresses that can be easily adjusted for firmness by the user. Sleepers can tailor the mattress to any firmness they desire, change the firmness as often as necessary, and even have different degrees of firmness for different sides of the bed. The concept of an "air bed," however, is new to most consumers, presenting a challenge in convincing them it is worth buying.

Select Comfort did not invent its technology; the concept had been in existence for some time when the company acquired the patents. It commissioned studies at major universities, including Stanford, the University of Memphis, and Duke, to show that their product enhanced back comfort and resulted in a more restful sleep (e.g., 24% more rapid eye movement, or REM, sleep).[15]

Select Comfort has launched an aggressive communications campaign focused on proving its product's core benefit, a good night's sleep. The company maintains its own chain of about 300 retail stores in high-traffic areas where consumers can take a "Test Rest on Air" and ask questions of highly trained representatives. It also formed an alliance with Bed Bath & Beyond, a leading home furnishings chain, to market its products through over a dozen of their stores. Select Comfort provides demonstration videos as well, especially useful for someone to take home to share with a fellow household decision maker, and provides savvy collateral material that discusses "Sleep Science" and the problems of back pain. Consumers can visit the company's Web site, which provides an organized, clear articulation of the products and their benefits, and even offers advice on sleeping better, such as herbal remedies for insomnia.

A mattress is a risky purchase; consumers pay on average around $1,200 for their mattresses and tend to keep them for 11 or 12 years. To reassure consumers, Select Comfort has secured endorsements from sports celebrities such as Cris Carter, a running back for the Minnesota Vikings, and well-known spokespersons such as the radio announcer Paul Harvey. It also relies on an active campaign to encourage customer referrals (customers are paid).

As it matures, Select Comfort faces a range of marketing issues, including extending its product line, broadening its focus beyond just "the air bed company," and increasing retail distribution efficiency.[16] But for now, the company has experienced considerable growth by taking a largely unknown product innovation and using sophisticated marketing and retailing practices to prove to the public it can deliver a valued benefit.

The Kodak Advantix APS camera is an example of a superior product that stumbled because of a failure to articulate benefits. The Advantix system, introduced in the mid-1990s, is a leapfrog in design over established systems by combining several high-tech features to ensure ease of use, guaranteed high picture quality, and flexibility in picture formats. The original product launch in 1995 was at the time the most costly ever in Kodak history, but it produced disappointing results. One of the major problems was a failure to clearly point out the product benefits to the public. Kodak changed advertising agencies, and in 1997 it launched a campaign orchestrated by Ogilvy & Mather that focused on the product advantages of easy drop-in loading, the availability of three different picture formats, and an index print feature that avoids the need to search through negatives when ordering reprints.[17]

Implementing a Strategy to Prove Benefits

Identifying benefits of a new technology requires a two-prong approach consisting of introspection and market feedback. The first thing a marketer should do is delineate the unique features of the new technology and explain why it is superior to existing technology. In doing so, the marketer must link each advantage with a statement of the core benefit, meaning the reason a buyer would be motivated to acquire it. This is not always easy, particularly for technical people, who are intuitively familiar with their product's superiority but are unaccustomed to communicating with skeptics who lack an interest in technical details. One of the authors consulted for a company that sold sophisticated Smart Card systems to large organizations. He helped their management clarify benefits using what we call the "napkin exercise," instructing them as follows:

> Pretend you are on a plane and find that by accident you are sitting next to the CEO of your largest

*prospect, a golden opportunity for selling. Further, he or she is interested in hearing about your product, and wants to give you a chance to pitch it right there. The problem is the plane is landing and you have only 10 minutes. Further, all you have is a paper napkin to write down the top selling points of your system. And, the person is completely **non-technical** and needs you to write down an explanation of why each feature **really matters** to their organization. Instructions: Write down the most important features of your system and the reason why each one makes a difference or adds value for the customer. Remember, you have to do it on the back of a napkin and cannot use technical jargon.*

The outcome of this introspection should be a clear statement of exactly why the technology makes a difference. For example, a Smart Card can store several credit card and bank account numbers and link them to a single universal number, all in one memory-enhanced card. The value of this feature is not that obvious, so it is important to explain that the card holders in the system can now carry around just a single card. Their wallets are no longer cluttered, they can manage their accounts better, and they are better protected from losing all their cards and checks when a wallet or purse is stolen.

Besides an introspective look, a marketer should also talk to consumers. Consumer feedback can be gathered through traditional market research methods, including focus groups, in-depth interviews with customers, and surveys. The format of the consumer dialogue should include presenting the creator's vision of the technology and attempting to learn (at a minimum) the following:

- How well do consumers understand the vision?

- Once educated about the technology, do consumers have their own vision, and what is that vision?

- What is the link between product/service features and how do they have a positive influence on the consumer's life? For example, does it save the consumer time, money, stress, pain? In what ways does it contribute pleasure, inspiration, excitement, peace of mind?

- In what language should benefits be articulated so consumers can understand them?

Finally, a company must rely on marketing communications to get its message out, investing in advertising, publicity, brochures, sales, and new media communications. The nature of the message will vary depending on whether the target is a consumer or an organizational buyer. Consumers can usually understand the benefits of technology in qualitative terms, but businesses need to see the impact on the bottom line.

An effective way of communicating with organizations is to formulate a thorough business case that compares the net benefit of the new technology to established products or services (or older technology). This kind of approach was used to help prove the benefits of photocopying compared to carbon copies and mimeographs. It was also used to demonstrate the value of local area networks in their early stages. Value is defined as follows:

> **NET BENEFIT** = *Net Present Value of Cash Flow*[18] *of [($ Benefits of New Technology − $ Benefits of Old Technology) − New Technology Acquisition Costs]. . . where the $ Benefits refers to cost savings and increased revenue.*[19]

Quantifying dollar benefits requires a broad approach, looking at the different ways the technology affects a business process. A good example is the introduction of a just-in-time (JIT) inventory management system. From a retailer's perspective, products would arrive closer to the time they are out of stock, reducing inventory requirements. Taking this into account, one could calculate the total savings from reductions in warehouse space rentals and lower interest from improved cash flow. One could also calculate savings from reductions in supplies and clerical time, since there is no longer the need to process paper orders. These savings would be added to the "benefit" side of the equation, along with revenue benefits such as increased sales. A marketer of such a system would quantify the financial savings and subtract the cost outlays of the new technology over a period of time. Finally, a return on investment would be calculated by discounting the cash flow, that is, taking into account the fact that a dollar today is worth more than a dollar tomorrow, and dividing this by the acquisition costs.

A thorough analysis of benefits for a business technology is time-consuming and may require gathering data from case studies or surveys. Another method is to implement a pilot test in which customers are asked to use technology and are subsequently debriefed about the impact it has on their business. To illustrate, the Smart Card is a promising technology that lags because it must gain simultaneous acceptance among consumers, merchants, and banks; for example, consumers only use the cards in commerce if they are widely accepted by merchants, while merchants do not want to accept them until a substantial share of consumers are using them. To overcome this problem, the financial services industry has sponsored a number of trials to prove the benefits, accompanied by extensive surveys to quantify the benefits among different parties. This included a trial during the Olympic Games in Atlanta sponsored by Visa, NationsBank, First Union, and Wachovia.[20]

If a technology is relatively new, a solid "business case" showing the returns from the technology may be the best way to prove its benefits and open the eyes of potentially skeptical buyers. Consumers may also benefit from a formal, dollars-and-cents business case, particularly when continual cash flows are involved, such as innovations that reduce energy costs (e.g., a solar heating panel). At the same time, it is also important to stress the intangible benefits of a technology, the ways in which it creates happiness and alleviates problems that are hard to quantify in monetary terms.

Whether relying on a structured approach that presents the monetary benefits of a technology, or savvy communications that demonstrate the power of a technology and how it can have an impact on a consumer's life, proving benefits is an essential strategy for marketing a cutting-edge technology. Explorers and Pioneers may be willing to experiment with technology and accept on faith that it is beneficial to them, but the attitude among Skeptics is "show me."

MARKET-STAGE PRICING

It is not uncommon for the price of new technologies to drop steadily after introduction to the market. This pattern is seen over and over; for example, prices dropped for automobiles, televisions, microwave ovens, mobile telephone service, and countless other products and services after they were first introduced. The first high-definition television sets introduced in early 1999 cost close to $10,000, but newer models are already becoming affordable. Picture phones and related videoconferencing technologies are continuing to drop in price, promising a revolutionary change in communication in the near future.

Price is an important variable in explaining the adoption curve for a new and innovative product or service. When forecasting demand, sophisticated analysts incorporate changes in price into their models to produce more realistic projections. It could be argued that price is an exogenous variable, an outside influence that marketers should respond to rather than actively incorporate as a key marketing mix decision. Thus, price might be dictated by production costs and other economies of scale (such as distribution or infrastructure), with providers continually passing on their savings and the market responding, in turn, with increased sales.

An extreme view would be that price is the dominant variable in explaining the adoption curve for a cutting-edge technology. It might be argued, for instance, that every American wanted an automobile from the onset of the twentieth century, and that it was primarily lower costs resulting from the shift to assembly-line production that led to the product's wide penetration. To illustrate, Ford priced its Model T at $850 in 1908 and sold 18,000 units, but in 1925, when it priced the model at $260, it sold 2 million units.[21] In effect, one can expect a natural drop in costs as production and distribution approach optimum levels, which is reflected in prices and, correspondingly, in consumer demand.

Our basic research on technology readiness suggests that as the market for a new technology matures, marketers have reasons to lower prices having nothing to do with production costs. Specifically, the remaining consumers who have not acquired the technology will be willing to pay progressively *less*. Indeed, lower production costs may also play a role in how prices are set for a given technology, but are not the only factor. Marketers need to base pricing decisions on consumer demand considerations as well as supply costs. Market-stage pricing is the strategy of adjusting pricing and product alternatives to meet the needs of consumers at different stages of a market's maturity.

Why would consumers become more price sensitive as a technology matures? Two factors influence the willingness to pay for a new technology: *perceived value* and *ability* to pay. As different segments of the market become ready to acquire a technology that has not reached its full potential, there is increasingly less optimism, or a general belief in the value of technology. In addition, the affluence of potential buyers tends

EXHIBIT 7-4

FACTORS INFLUENCING PRICE SENSITIVITY: OPTIMISM AND INFLUENCE

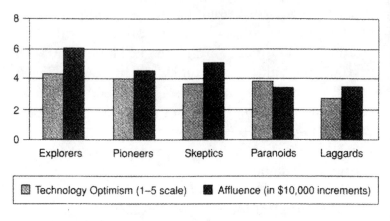

to be less than among segments who entered the market earlier.

Exhibit 7-4 depicts segments from our technology consumer typology, arrayed from the highest to the lowest in overall technology readiness, which is a proxy for how quickly they will acquire a techno-product or service. The exhibit shows the segments' overall optimism toward technology and annual household income. Explorers have the highest optimism and income; levels for both variables drop progressively as segments become less techno-ready, with Laggards having the lowest optimism and income (the pattern is not perfect, with Pioneers tending to earn less than Skeptics but more than Paranoids and Laggards). As a market matures and becomes dominated by less-techno-ready segments, the pressure to drop prices is likely to increase because consumers care less and can afford less.

A business-to-business marketer can expect similar reactions to price as a technology matures, although more research is needed in this area. More-techno-ready corporate decision-makers will more readily recognize the benefits of cutting-edge technology and make a more forceful business case for acquisition. We also find, when segmenting businesses based on benefits sought, that more-techno-savvy businesses are often larger and more profitable, and can therefore afford the latest technology.

Given the evidence that price sensitivity will change over time, what are the implications for technology marketers? Certainly, there is no rule that states one must charge a high price in the beginning and drop prices steadily over the long run. The strategy of market-stage pricing, however, implies that marketers need to pay attention to the stage of maturity of their technology. Because the newer customers tend to be less techno-ready, marketers need to respond, with tactics such as passing on savings from production economies, creating more basic models that are affordable, or offering promotions that give more price-sensitive consumers an opportunity to buy. For example, Microsoft offers four versions of its Office 2000 suite of business applications: Standard, Small Business, Professional, and Premium. Each comes with different features, the Standard and Small Business versions having the fewest applications but costing 29% to 45% less. The consumers and small businesses that buy these more basic suites are likely to be less techno-ready, less demanding in terms of applications, and less willing to pay higher prices. Software companies frequently rely on special versions or bundling to optimize their revenue in a market where there is a wide disparity in price sensitivity.

As a general approach, marketers should continually evaluate their pricing as their technology offering becomes more commonplace, but the real world is never simple. Pricing decisions may depend on a variety of factors besides changing consumer price sensitivity and costs. The following are situations where a technology marketer may consider a strategy different from starting out with a high price and steadily lowering it:

- **The competitive nature of the market.** Is the offering truly unique relative to competition, or are there several companies introducing their own versions of the same technology at the same time? For instance, cable modems initially offered unparalleled benefits of high-speed access, making it possible to skim a substantial premium from its first users. The increasing availability of high-speed DSL lines to homes and businesses creates a more competitive market that will exert price pressures on cable systems for high-speed access.

- **The trade-off between long- and short-term profits.** Will accelerated sales result in greater profit in the long run (despite lower profit now)? A company may deliberately charge lower prices to expedite demand and achieve critical mass in the market. This model is used by many dot-com companies in today's market and explains why many continue to run in the red despite growing sales. Examples include online retailers targeting a product niche, such as Amazon.com.

- **The desire for long-term relationships.** Should prices be discounted in order to "buy" a new customer? This is particularly applicable to a high-tech *service*, which depends on streams of revenue from a base of customers. In an attempt to sign up people using the technology for the first time, and hopefully locking them into a long-term relationship, a firm may decide to price lower than necessary to break even. The price may result in a unit loss, or it may result in lower dollar sales in exchange for higher unit sales. A wise marketer using this strategy will stipulate a fixed commitment to ensure that the new customer does not switch to a different service. For example, a highly successful regional Internet provider, Erols, offers a discount in exchange for paying for a minimal service term in advance. Cellular telephone providers have offered promotional prices to sign up new subscribers, but have stipulated a minimum service period as part of the contract.

- **The revenue model.** What impact does the price charged to one particular customer group have on the overall strategy for earning revenue, which may include sales to other types of customers? Every technology marketer should formulate a *revenue model* consisting of a plan for extracting sales from multiple sources. For example, many new media and e-commerce ventures struggle to determine the optimum strategy for generating revenue, considering sources such as subscription fees, fees for using add-on services, advertising sales, online space rental, commissions, fees for referring visitors to other sites and the sale of user data. An entertainment site may provide free access and attempt to make money from advertising sales, or it may opt to charge for access to earn revenue. In a complex economy, companies find they have a wide range of options for earning money, of which unit sales are just one. Depending on its planned model for earning revenue, a technology marketer may choose to charge one group of customers a reduced price or even offer it for free, hoping to bolster sources of revenue from another source, such as advertising.

- **The message that price sends about quality and risk.** Given that a revolutionary product or service is an unknown, will consumers believe it can safely meet its promises if the price is unrealistically low? It has long been known that buyers may use price as information, especially when they lack other kinds of objective information for a decision. Too low a price could signal that there is problem with quality, a hidden catch, or even a risk (say, a fly-by-night company or a scam).

For the marketer of technology, there is a need to continually respond to the inevitable flux in price sensitivity as the stage and composition of the market changes. As a general rule, the marketer should plan for *decreasing price thresholds,* since new customers tend to be less techno-ready, a situation that can be met with the introduction of more basic product offerings or passing on savings from greater efficiency. In addition to this general rule, the marketer should consider the complex dynamics of technology markets in establishing a pricing strategy. A company may decide its products are so unique they can be premium priced, or it may decide it faces competitive pressures from a new generation of technology, forcing prices down. A decision might be made to start with a low price, even below cost, in the interest of achieving an invincible position in the market or locking in long-term relationships. It might be deemed wise to keep unit pricing low or even free in the interest of producing revenue from other sources, such as advertising or add-on sales. And finally, a company may decide to deliberately price its technology on the high side, thereby creating a perception of superior quality and less risk.

Pricing techno-products and services is a complex process that requires careful consideration of the long-term implications. Technology usually shows a trend to decreasing prices, a phenomenon we believe is driven as much by consumer tastes as by economies of scale. It is essential for marketers to pay attention to the stage of maturity of the market for their technology and the implications it has for pricing. It is also important, however, to rely on thoughtful strategy as well as instinct, perhaps choosing to ignore a natural downward progression in favor of a more calculated decision that helps a company win in the end.

ACQUIRING TECHNOLOGY CUSTOMERS

Marketers of techno-driven products and services face two major challenges because of the unique consumer behavior surrounding technology: (1) acquiring new customers and (2) satisfying existing customers. This chapter has dealt with the issues of acquiring customers, recommending strategies suited to innovative technology, which are reinforced by our research on consumer beliefs. Having placed these acquisition strategies in perspective, the next step is to categorize them in the order they should take place.

- *Technology evangelism* and *future-ready design* should occur in the earliest stages of development of a technology market. Technology evangelism seeds the market by instilling early adopters with enthusiasm, helping to create a momentum that propels sales as the market expands. During this expansion stage, future-ready design makes a company competitive because its own techno-offering includes features that increasing numbers of buyers will discover to be valuable.

- There is no reason marketers should not address the issue of *proving benefits* early in the development of a technology market, but this strategy becomes *critical* as the market approaches maturity, because increasing numbers of unsold consumers will be skeptical.

- *Market-stage pricing,* by definition, should be practiced throughout the entire life cycle of a technology until it ceases to be innovative. Marketers need to continually gather information on consumer price sensitivity and respond with creative strategies when new buyers appear more value-oriented. Technology marketers must also continually evaluate the impact of their pricing decisions against the backdrop of other strategic concerns.

The idiosyncratic nature of marketing technology should continue long after the sale is made. Satisfying and retaining hard-won customers is an equally challenging task when the product or service is techno-driven. The next chapter shares some additional research we have conducted on how consumers evaluate technology and provides strategies for managing relationships.

CHAPTER 8

Satisfying
The Technology
Customer

The third principle of techno-ready marketing holds that technology marketers face special challenges in ensuring customer satisfaction. A techno-based product or service is likely to be cutting edge and hence unfamiliar to many of its users. Furthermore, it may remove much of the human component found in earlier solutions to the same problem, taking away opportunities to appeal to another person for help. But the greatest challenge of all is the heterogeneity of the market for a techno-offering; many consumers who feel the need to acquire a new technology will possess a high level of discomfort in using it, while others will feel insecure about it working safely and properly.

The problems of making customers feel comfortable

and assured about technology increase as a category matures and a larger share of the population chooses to adopt it. The first adopters—the Explorers—are bold, independent self-learners, but those who follow will have decreasing tolerance for technology systems that come off as complex and uncooperative. As we discussed in a previous chapter, marketers who address these consumer issues early on can gain in many ways. Their customers will be more satisfied and their brands' market shares will grow more rapidly. In addition, techno-based products and services that are engineered with the needs of the least techno-ready customers in mind will be easier and more cost-effective to support.

The solution is to focus on the low-TR consumer to learn how to design, support, and talk about a technology. Just as the high-TR consumer tells marketers how to make techno-driven offerings *appealing,* the low-TR consumer tells marketers how to make them *accessible.* Customer feedback is critical, since the managers and engineers empowered with designing and planning the implementation of techno-offerings are likely to be tech-savvy. Highly techno-ready managers will find it hard to understand the difficulties faced by so many of their customers who are eager to buy but are restrained by discomfort and insecurity. The techno-savvy marketer *must* understand. While the belief set of these less-techno-ready customers may make them seem handicapped, the technology marketer must address their concerns because they comprise half the market. Some people load software and figure out how to use it by navigating through menus; others need to take a course. Some take a product out of a box, assemble it, and start using it; others get someone else to assemble it and show them how to use it. Some know and trust that technology is working; others need a sign, a reassuring message that things happened the way they should have.

Understanding must also be coupled with humility. The low-TR individual is not necessarily less capable, since technology readiness is a mind-set, not a competency. High-TR people may solve a problem with self-help, not realizing they spent hours tinkering around, while low-TR people pride themselves on solving a problem with a quick phone call. High-TR people may feel the dangers of Internet commerce are grossly overstated, while some of the low-TR persons' paranoia may be validated by security experts.[1]

We have investigated the issue of customer satisfaction for techno-driven products and services in our ongoing research on technology readiness. In one study among college students and young professionals, we quantified the perceived quality of Internet service providers (ISPs) along several dimensions relating to reliability, responsiveness, empathy, assurance, tangibles, and security.[2] For each dimension, we compared the ISP's perceived *performance* with customers' *expectations* by using the SERVQUAL scale, a tool for assessing quality for services.[3] We were also able to relate users' perceptions to their technology-readiness levels, as measured with an early version of our TR Index.

Internet access at the time of the study in early 1998 was an unreliable service that often failed outright. A large share of users in the study perceived their service as falling below their least acceptable performance levels in areas ranging from getting connected to accuracy of transactions. Consumers who were low in technology readiness clearly showed more disappointment with their service, with levels that were frequently appalling (up to half perceiving outright failure). These low-TR consumers more often perceived that their ISP failed to

- Ensure that their transactions were secure

- Have their best interests at heart

- Ensure easy installation

- Ensure that the system would never go down

- Have courteous employees

- Offer help when needed.

More important, the specific TR traits that drove this perception were *discomfort* and *insecurity*. The motivating aspects of TR, optimism and innovativeness, bore no relationship in the research to service perceptions. The conclusions are as follows:

- The inhibiting aspects of technology beliefs play a role in the perception of quality of a technology-driven service (our broad experience in consulting suggests this applies to products as well).

- The impact is negative, not positive, contributing primarily to perceptions of failure.

- The impact is widespread, affecting perceptions of security, reliability, and even the belief that somebody cares.

- To the extent that failure to satisfy needs is related to behavior, negative beliefs about technology can play a role in holding back usage and adoption of technology brands. The obvious case is when the user has trouble with installation, a problem that could easily lead to switching to a more user-friendly option.

Meeting the needs of all customers, not just the most tech-savvy, does not have to be an accident. Many companies have an extensive dialogue with their customers, heed the advice of the less techno-ready, and engage in practices to ensure that technology is problem free. The following discusses three particular strategies that relate to product design, servicing, and communication:

- Customer-focused design

- Responsive customer care

- Reassuring communication

CUSTOMER-FOCUSED DESIGN

The process of ensuring that technology is responsive to all types of customers, not just the techno-savvy, begins in the design phase. Technology must be customer-focused; this means it can be used by all who choose to adopt it, that problems are minimal even for those least comfortable with technology, and that the method of operating and controlling does not detract from obtaining value. By being customer-focused, technology goes beyond just delivering the right functionality; it takes into account human factors.

Customer-focused design specifically addresses the needs of the consumer who has a high degree of technology discomfort. The primary issue for such people is *control.* Uncomfortable consumers feel they lack control over technology; they tend to believe technology is complicated, hard to work, prone to failure, and often designed for elitists. Even technology optimists—who have a general faith in the value of technology—can feel this way. Thus, these optimistic but uncomfortable consumers may attempt to use technology despite the challenges involved, and when given options, seek out brands that are easier to use.

The concept of customer-focus applies to any techno-offering; it is important in the control features for electronic hardware, the navigational aspects of software and Web sites, and the directions for using a techno-service such as an ATM. Customer-focused technology does not have to be patently simple, just not overly complex. Too much simplicity could result in limiting the functions that might be performed, and could create unintended frustrations such as slowing down the speed of use. The designer must strike the right balance so that all users are comfortable with the technology and feel in control.

If designed to be customer-focused, technology will have the following characteristics:

- **Intuitive:** The protocols for making the technology perform as desired tend to match what people would expect, and as such, there is maximum potential for learning it without seeking help. For example, the labels on buttons on an appliance, or the links on a Web site, lack ambiguity. To be intuitive, controls must be easy to find and read. Careful attention is paid to location, size, hue, and feel. Web designers are concerned not only with the aesthetics of their design but also the ability of the average user to clearly view and find content.

- **Efficient:** It takes minimal time to complete a task using the technology. Controls are easy to find and operate. Also, controls that involve a sequence of steps, such as moving through layers in a menu, are grouped logically and efficiently; this means there is a good balance between the number of options to choose from during each step of operation and the number of times a choice must be input. A good example is an automated voice response menu for a telephone service: too many choices are cumbersome, but if the options are reduced too much, the customer must make selections too many times to find the desired option. (This problem is mirrored in the trade-off between page size and layers in a Web site.) Efficient technology also minimizes unnecessary repetition, avoiding the need to conduct two tasks to activate a feature when one would suffice.

- **Responsive:** It takes minimal time for the technology to respond to a task. A common issue in Web-site design is the speed of downloading of pages; for instance, sites designed for users who prefer a simple appearance should avoid excessive graphics content. Software that stores and manipulates data, such as financial applications, often include options that avoid slowing down the process, such as a "batch" mode for entering data.

- **Assuring:** The technology offers cues that it is working properly and provides useful feedback when there is a failure. An example that is so basic it is taken for granted is the dial tone in a telephone, which signals that the user is connected to the network.

- **Compatible:** The technology fits seamlessly with other technology the user will possess. The customer should not find it necessary to rig the technology or take special steps to make it work. If technology is designed for a mainstream market, it should mesh with mainstream systems. This is a big issue in computing. To illustrate, computer game producers need to decide which CPU speed, memory, drive speed, and operating system their software will work with. There is a trade-off in that software designed for newer, more advanced computers will result in a superior experience for those who have them, while creating frustrations for those who do not. The onus is on the producer to target the software for the most likely users: are they involved game players who seek the most up-to-date features, or grade-school children using it for educational purposes on donated computers at school? The problem applies to a wide range of technology; for instance, advanced televisions and VCRs should be compatible with the feed of local cable systems, and Web sites should work with the typical visitor's browser.

- **Reliable.** The technology is dependable and free of glitches that would challenge a person with a low comfort level. "Bugs" are acceptable for a segment of early adopters, who enjoy the challenges of testing and overcoming a product's limitations, but can be crushing for a consumer with a low comfort level.

Case Studies in Customer-Focused Design

There are cases where a strong emphasis on customer-focused design can be the very essence of success for a product. In the 1970s, Polaroid introduced a low-priced instant camera called the One Step. True to its name, it required only one step to operate, which consisted of pushing a button once it was aimed at the subject. Previous instant cameras were complicated to operate, involving a series of cumbersome steps. The film for the One Step came in a cartridge and could be easily inserted in the camera with no need to open the cover or pull tabs. When a user snapped a picture, it ejected the photo automatically from the front of the camera. Each picture had a self-contained packet of chemicals that finished the image without requiring the user to time the process or peel paper as with older technology.

The One Step became an instant market success. Anyone could operate it, from a young child to a senior citizen. But, Polaroid and its retailers wanted to do more than just sell cameras; they wanted to sell the custom film cartridges used in the camera. Because a user could easily snap off a full cartridge in minutes, the camera had a voracious appetite for film.

At the time the One Step was introduced, the main benefit of instant photography was instant gratification, since users did not have to send photos out for developing. Polaroid recognized that the average person was uncomfortable learning and conducting multiple tasks to operate a camera. No matter how simple the steps were to many people, a large segment of the population found them to be either annoying or impossible. The One Step made instant photography accessible to a larger market.

Another classic case of customer-focused design is the graphical interface for personal computers. Up until the mid-1980s, computer software was controlled by two common interfaces, or "front end" systems: one was the command-driven interface, in which the user had to type instructions; the other was the menu-based interface, in which the user could select options from a menu, usually by typing a letter, scrolling, or clicking with a mouse. The graphical interface, common to millions of PC users today, represents choices with visual icons that, when clicked, open windows containing working areas for an application. The philosophy behind the graphical interface is that people can learn and respond better and faster to images than to commands, making computing a friendlier, more intuitive experience.

The first graphical interface was designed by Xerox in the 1970s, but the first widespread application came with the introduction of the popular Apple Macintosh in 1984. Following advancements in CPU power and monitor quality, Windows 3.0 introduced a graphical interface to PC platforms in 1990.[4] Today, it is a standard feature in all computer software. Computer users are able to accomplish more tasks, with greater efficiency and less learning than when they had to remember cryptic commands or scroll through cumbersome menus. In the future, users will be able to control their computers by voice, eye motion, and brain waves patterns.

A more recent example of customer-focused design is the iMac personal computer by Apple, a popular product that was part of the company's efforts in 1998 to make a comeback during a period of financial distress.[5] Apple realized that computer users were frustrated by the complex array of wires and plugs that make a personal computer such a problem to install, operate, and transport. Indeed, the personal computer has been compared to the television to highlight design inadequacies that have slowed its growth. The first televisions were self-contained units that were simple to operate, whereas even modern computers require a certain degree of training and technical ability to operate. The iMac is a paragon of customer-focused design, eliminating the spaghetti wiring by combining all functions into a single, compact, aesthetically appealing unit. The design is reassuring to consumer and business users who are frustrated with traditional PC setups.

Iridium, a mobile phone service that provided global satellite coverage, is an example of a technology concept that encountered marketing problems from its own complexity. The name was derived from the element with the atomic number 77, the originally planned quantity of satellites from which it would operate. The catchy name contributed to its image as an advanced, visionary technology, and it told the buyer that it was not a simple gizmo for simple people. The service encountered problems in reaching growth targets, which could be attributed to many factors: the unexpected global growth of inexpensive land-based cellular phone service, which competed with Iridium; large introductory prices and confusion over pricing structure; and problems supplying sufficient quantities of phone units to customers. But one of the major criticisms of the service was its strategy of targeting a mass market of business users. These consumers, who might be the same ones who would buy a phone at Radio Shack, found the technology not only expensive, but intricate and perplexing to use. Users required training to operate the service. In its introductory phase, Iridium also challenged them with reliability problems such as blocked access, interference, and dropped calls. In the end, the company changed its strategy to target large, multinational companies and niche markets. Iridium was testimony to paying too much attention to being a technological marvel while ignoring the more pragmatic requirements and comfort issues of the target market.[6]

Customer-focused design is not a new concept, nor does it occur by luck. Even early technologies have taken into account the human element. Samuel F. B. Morse and Alfred Vail labored over the creation of the first telegraph codes, making decisions such as reserving the simplest dot as the symbol for *e*, the most-often-used English letter.[7] Christopher Latham Sholes reworked the keys on the first typewriter in order to improve product performance, creating the now standard "QWERTY" format.[8] In the 1940s, the Bell System pioneered the development of the Touch-Tone phone as a method for speeding up dialing. Before the technology was introduced for home telephones in the 1960s, AT&T conducted extensive human factors testing to determine the precise position of buttons to increase dialing speed and reduce the rate of human errors.[9]

The Process of Testing and Refinement

Customer-focused design results from a concerted effort to properly define and understand the target market, design interfaces to ensure maximum comfort, and rigorously test the interfaces with potential users. A critical requirement of the testing is to include a broad spectrum of customers, including a segment who are low in technology readiness and possess a high degree of discomfort. The testing stage is important because it is difficult to anticipate in advance exactly which design features are easiest for consumers.

Marketers need to engage in some form of *usability testing*, a process for evaluating how consumers operate technology and for identifying improvements to make it more customer-focused. Usability testing is its own discipline, which pays close attention to the human factors of design, engaging in a structured process of feedback and measurement for techno-driven products and services. It is an essential step for technologies that are new (and therefore unfamiliar) or complex. The boom in computer software and the World Wide Web has spurred interest in usability testing, as well as an accompanying growth in specialized testing methodologies and laboratories.

Different approaches can be used for evaluating technology. Users might be shipped or asked to download products from the Internet in order to test them and relay feedback. Evaluations might be conducted in a focus-group setting or on a one-on-one basis. One approach is to have a user work at a computer under the direction of an interviewer. The interview and the computer activity may be videotaped, and product managers and designers may watch from behind a one-way mirror. A structured approach is often used in which the user is asked to complete tasks (e.g., "Find information on pricing"). The tasks are timed and the user is queried about what made the task easy or difficult and what might be improved.

One of our clients was upgrading a sophisticated software system for use by a select group of large, corporate customers for an information-processing activity. The client covered the expense of having supervisory personnel from its largest customers flown to a central location where an interview team provided a step-by-step walk-through of the latest version of the software. An advantage of having the supervisors rather than hands-on users view the prouct was that they had a keen ability to identify ways of streamlining and simplifying. Although they were not day-to-day users, they were responsible for training and overseeing those who were. The supervisors had a stake in increasing efficiency in using the software and in reducing the amount of time spent on training. The meetings also provided an opportunity to ask other relevant questions to make installation easier, such as the types of computer systems and software already in place in different customer shops.

In deciding which route to pursue in evaluating technology, a marketer must have a clear idea of informational goals. Usability testing that is highly structured is not very effective in addressing strategic issues such as how design influences the perceptions and usage patterns of the technology. For example, in usability testing for an e-commerce site for travel, a user might be instructed to make a seat selection; the process would reveal how to make the task as clear, simple, and efficient as possible. But the general design of the site may be flawed in that it fails to communicate that the seat-selection feature is even possible. A less-structured approach, such as a navigational tour in a focus-group format, might be more effective in identifying these broader issues. Marketers should begin with more general research to address functionality, positioning, and broad navigational issues with technology. After creating and refining a prototype, it then makes sense to apply more focused usability testing.

Customer-focused design ensures a more satisfied customer and makes a techno-driven product or service easier to sell. Those who benefit the most are the customers with the greatest technology discomfort, and it is critical to include these customers in the evaluation process. If properly executed, a customer-focused technology will be responsive to all customers, from the most uncomfortable to the most techno-ready.

RESPONSIVE CUSTOMER CARE

Even with the most customer-focused design, almost every user will on occasion encounter a problem with technology. The instinct for solving problems will vary from one user to another and will depend greatly on the level of techno-readiness. The highly techno-ready consumer will instinctively try to solve the problem alone. These consumers are independent self-learners and are more often in the role of helping others than seeking help. Because their comfort level with technology is high, they will attempt to trouble-shoot before they resort to looking something up or calling an expert.

The low-techno-ready consumer is not as self-reliant. The technology may seem incomprehensible, a puzzle that only the sophisticated user can solve. Again and again, research shows that low-TR people would most prefer to seek help close by, turning to a friend, relative, or coworker who is viewed as more capable and familiar with the technology. This sometimes creates an odd reversal of roles, the parent dependent on the 12-year-old child or the executive in awe of the college intern. An expert is not always available in the next room, however, so the instinct of the consumer with high discomfort is to pick up the telephone to talk to a live person. The experience is not always positive. The National Technology Readiness Survey shows that two-thirds of low-TR adults believe that "when you get technical support from a provider of a high-tech product or service, you sometimes feel as if you are being taken advantage of by someone who knows more than you." Among low-TR customers, 69% agree with this statement (compared with only 27% of high-TR adults). Consumers are often frustrated by the perception of being patronized, deluged with meaningless techno-babble, or given the runaround by poorly trained representatives.

Exactly what kinds of support do consumers require for techno-driven products and services? The NTRS asks consumers how desirable they find different approaches to technical support, as shown in Exhibit 8–1. Con-

EXHIBIT 8–1

PREFERRED METHODS OF TECH SUPPORT

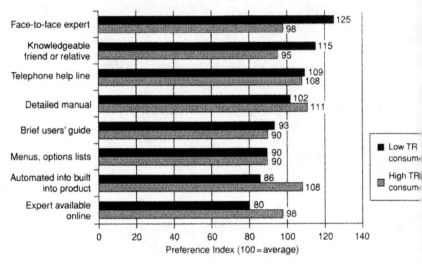

Preference Index (100 = average)

sumers with high levels of technology discomfort tend to find most forms of technical help to be less desirable than their more techno-savvy peers. For this reason, the desirability ratings reported here are standardized to reflect their desirability relative to each other, conveying a sense of preference, or in the case of the more uncomfortable consumer, the degree of being a lesser evil. Consumers indicate they want the following kinds of help:

- Consumers with a high degree of discomfort with technology prefer to deal with an expert face-to-face, someone who comes to their home or office or is accessible at a store. Their second choice is a friend or relative who is an expert, and their third choice is to communicate with technical support by telephone.

- Consumers who are comfortable with technology most prefer a detailed manual that shows how to use, and solve problems with, the high-tech product or service setup. Their other favorite choices include a telephone support line and automated information that is built into the product or service that can be accessed when needed (e.g., "wizards," which are common with a lot of software).

- Online help is good for the more tech-savvy, but a problem for the less comfortable consumers who are less likely to be online anyway.

- A built-in self-help feature, while great for those who are comfortable with technology, is among the least desirable options for those with technology discomfort.

- There is general agreement that the least useful form of support is from more limited help features, whether they are built into the technology in the form of menus or option lists, or accompanying it in the form of a brief users' guide.

The preferences highlight differences in problem-solving styles among consumers. Those most comfortable with technology wish to be empowered to help themselves. Those least comfortable instinctively turn to people, preferring an expert they can trust.

The corporate decision about how to support customers is often based on economics. Companies may prefer built-in self-help features because they are cost-effective. Although potentially expensive to develop up front, they may be offered as a primary source of help in order to forestall the high costs of printing manuals and staffing telephone help lines. If telephone support is available, a company may decide to charge for it, or may limit its accessibility by confining it to a narrow time window, or staffing it so sparsely that users encounter long wait times. The irony of offering self-help features in technology is that they are most effective for the highly techno-ready user, the one who least needs help of any kind.

Not all companies will limit the ways their customers can obtain support. Those that want to experience continued growth beyond the introductory stages of a technology will ensure that all customers have the kind of help they need. Such a company is likely to establish a customer service function that is adequately staffed and trained to serve the needs of a wide range of sophistication levels. It will also consider other forms of support, including detailed documentation and perhaps even face-to-face support.

A strategy of intensive customer service may reap rewards beyond satisfying customers, and it does not have to be terribly expensive. Companies may leverage the service support function as a feedback mechanism for improving the product or service, since customer inquiries serve as a rich data source for identifying and correcting glitches. Steps can also be taken to improve the efficiency of customer support while empowering customers to solve their own problems. For example, companies that offer technology systems, such as an ISP or telecommunications service, can reduce the need for personal contact during outages by offering a recording stating the cause of the problem and the estimated time of restoration.

Models for Customer Support

Personal customer support can be offered in many forms. Sending an expert to meet the customer face to face may seem expensive, but it is usually feasible for a big-ticket business application. For instance, a telecommunications service may provide an on-site representative for a large account. Another model is for a company to train its customers to provide their own in-house service support.

It is not out of the question to provide a consumer with personal tech support, and an efficient way of achieving this is through a storefront. Gateway 2000 Inc., the successful national computer seller, began its business by selling exclusively from a direct-business model relying on telephone, fax, and the Internet. However, the company deals with a mass market that includes many new and amateur computer users who prefer a more personal touch. It established hundreds of Gateway Country storefronts, which feature its popular computer systems and provide trained representatives who can discuss needs with customers.[10] Its locations do not actually offer merchandise; customers still have to order their equipment and have it shipped, but the Gateway locations provide a face, which means a lot to those who are uncomfortable with technology. Another company that followed a similar model is Erols, a regional Internet service provider based in the Washington, D. C., area. Unlike the direct marketing models of its larger competitors, Erols opened stores in its trade area and signed up customers in person. Their Web site invites customers to "come surf the Net at one of our convenient mall locations."[11]

In our modern world, the telephone customer service center is a must for any high-tech product or service. Many companies prefer to concentrate on their core competency of designing and producing technology, leaving the job of customer care and technical support to a specialist. One such specialist is NEW Customer Service Companies, Inc., a Dulles, Virginia, provider of product protection and information services. NEW maintains state-of-the-art Customer Care Centers, which are open 24 hours a day, seven days a week. Their system supports 1,200 agents who handle up to 50,000 calls an hour. The company's clients come from a wide spectrum of industries, including consumer products, computer hardware, online services, and retail and financial services.

One of the services offered by NEW is a computer help desk. According to their Web site, "We've discovered that when a new purchaser of a computer returns it to the store because 'it doesn't work,' there rarely is anything wrong with the unit. Most computers come pre-loaded with a lot of advanced software and that's where the problems really occur. Our Computer Help Desk Representatives are ready to assist with any problem the consumer is having—software or hardware. With our extensive software library and hardware resource center, we're able to take a frustrated and unhappy customer who is ready to return the computer to the store, and transform them into someone excited by their purchase and grateful that the retailer they purchased the computer from was able to provide this service."[12]

As more commerce gravitates to the Web, an increasing amount of customer service and technical support will be provided online through e-mail or an interactive chat with an "e-rep." Even consumers with a low degree of techno-readiness will seek out help online when their question or problem occurs while using the Internet. Recognizing the movement of service to the online medium, NEW offers its own branded service, Repairnow.com™. Another company that has plunged into the online service industry is PeopleSupport.com, based in the Los Angeles area, and includes among its clients leading online retailers. According to the president, Lance Rosenzweig, 66% of all consumers who start a shopping cart on an e-commerce site end up abandoning the transaction, a problem that may be attributed to customer frustration with online technology. Transactions can be smoothed, and more visits can be converted to sales, if the customer has access to real-time online support. Customers are given a choice of talking to an e-rep online immediately, sending a question by e-mail, or calling a phone number (70% choose to talk to the e-rep, who comes online to answer questions and guide the customer through a purchase).[13]

An interesting aspect of PeopleSupport.com is the emergence of an e-culture that is distinct from a traditional call center culture. Reps tend to be more oriented to and adept at written communication, better educated, and more Internet-savvy. Answers to many of the standard questions customers have are captured in canned responses that can be inserted into a chat automatically. Taking advantage of the chat environment, e-reps will actually converse with more than one customer at a time, increasing their efficiency over that of a phone representative.

Failure to provide responsive customer service and technical support can sometimes lead to a disaster. One of the authors worked with a company that introduced a techno-service without providing fully staffed technical support. The service itself failed to adhere to the principles of customer-focused design in its installation setup, so the average customer usually needed assistance in getting started. Only a dozen technical support representatives were ready in the "pit" to answer questions when tens of thousands of people attempted to sign up for the service. The result was that the majority of customers who requested start-up kits never actually activated the service. They tried to install their software, were thwarted, could not get through to technical support, and ended up abandoning the service in favor of a competitor.

Managing the Customer Care Function

A major issue in delivering customer support for techno-driven products and services is the technology readiness of the people who communicate with customers. This is an area that perplexes many managers and needs to be studied further. Consider the following issues:

- Will highly techno-ready customers be frustrated or insulted if the representatives treat them as if they were novices?

- Will highly techno-ready representatives become frustrated with customers who just don't seem to understand them?

- Should a company hire people with natural empathy and people skills and train them in technology, or should it seek out people with a natural interest in technology and train them to be effective communicators?

These are issues that need to be examined in ongoing research. Our technology-readiness research provides one clue that is encouraging: highly techno-ready employees who are likely to adopt technology also like to (1) learn about the subject and (2) tell others about technology. This suggests that the ideal service rep should be high in technology readiness. The National Technology Readiness Survey reveals that 46% of workers would find it desirable to "provide advice and help to customers or clients on using technology," but 69% of those who are highly techno-ready would find this activity desirable.

We have found it useful to talk to representatives themselves to better understand and benefit from their perspectives, focusing in particular on those who are considered to be the most effective in their jobs. In talking to these individuals, it is clear that the most effective technical support staff have a versatile style. They can ask probing questions and explain solutions in plain language to customers who are not comfortable with technology, while engaging as a peer with the more techno-ready customer. A valuable skill is being able to ascertain early in the conversation, usually by asking probing questions, just how techno-savvy the customer is. Reps tell us that one of the most difficult kinds of customers to handle are the "know-it-alls," people who lack hard knowledge about technology but perceive themselves as being experts.

A technical support environment can be different from a traditional call center, since the problems are more complex and the workforce is often higher paid and more educated. One of the factors that contributes to success, aside from careful recruiting and training, is the creation of a collegial atmosphere. Staff will learn from one another how to handle challenging problems and how to develop flexible communications styles.

A common decision that managers of customer care centers need to make is the procedure for follow-through. When technology is involved, there may be a higher priority to verify that customers were successful in solving their problems by contacting them after a conversation. The complexity of technology may also make it more important to have customers contact the same person for an unresolved problem rather than starting with someone new who is unfamiliar with the issue. The organization of customer service will ultimately depend on the individual situation—the nature of the technology, the techno-readiness of the typical customers, and the resources the company is willing to commit to satisfying customer need.

The issue of consumer discomfort with technology lingers long after the acquisition of the techno-product or service, making it imperative to provide responsive customer service and technical support. Faced with the decision of what kinds of help to provide, companies need to recognize that the customer with high technology discomfort has special needs. These needs often include access to a real person, either in a face-to-face environment, on the phone, or perhaps through the Internet. The cost of responsive customer care may be high, but companies must ultimately weigh this against the long-term desire to satisfy customer needs and compete effectively.

REASSURING COMMUNICATIONS

Insecurity is a major inhibitor to adopting technology, and it may have many facets. At one level, an insecure consumer is in greater need of assurance that technology is working properly. This can be illustrated by two scenarios. In one, a consumer calls up a catalog to order merchandise and is told by the representative that it will arrive in two weeks. In the other, a consumer goes to a Web site, orders merchandise, and receives a computer-generated message that it will arrive in two weeks. The insecure consumer will feel uncertainty about the online transaction. The human voice provides a sense of connection and a perception of ownership, whereas the online transaction seems to go into a black hole, the computerized promise seeming insincere. The more secure techno-ready consumer would tend to perceive little difference between the two transactions; a promise from a computer is as good as, or better than, one from a person (the person could make a mistake processing the order or make a promise that is unrealistic).

The ultimate level of insecurity is the feeling that technology is unsafe, posing a threat to one's social, financial, or even physical well-being. Referring to the two scenarios for purchasing merchandise, the consumer in each case may have provided a credit card number. The consumer who is insecure about technology would have greater faith in the telephone transaction. The process is familiar and the person on the phone is accountable. This consumer would be armed with stories about hackers stealing credit card information off Web sites. The more secure techno-ready consumer would argue that there is no difference at all between the two purchasing approaches, especially if the e-commerce site is "secure." In fact, this consumer might feel safer cutting out a person from the process since human beings can be dishonest.

Another form of insecurity relates to the risk of the technology. An insecure consumer is more likely to feel that, because it is new or eliminates some of the people input, it is more likely to fail. The more secure consumer would tend to disagree, feeling that newer, more automated technologies tend to be more reliable. The bottom line to the insecure consumer is that the time and money spent on a new technology could be a total waste.

What can marketers do to address these insecurity issues? We propose the following tactics:

1. Conduct a compelling demonstration about safety.
2. Provide reassuring features and messages.
3. Provide reassuring feedback.
4. Build a reputable brand name or leverage an existing brand name.
5. Provide guarantees.

Safety Demonstrations

Unfortunately, when a new technology is introduced, the accompanying publicity and firsthand experiences often exacerbate feelings of insecurity. In the world of the Internet, online service users receive instant messages from hustlers asking them to give out their passwords. An article on private espionage talks about how easy it is to intercept personal information from cable modems, which are increasingly common in homes and businesses.[14] Internet users warily turn on their computers in the new millennium wondering if their memory will be devoured by terrible viruses. Web browsers spew out alarming messages warning users to verify the integrity of sites when they download information, or asking them if they will accept a "cookie."

These issues have occurred for new technology throughout history, and marketers of the past have exercised creative ways to assure users. A case study in overcoming a consumer insecurity about technology is the introduction of early alternating current (AC) electric power systems by Westinghouse in the 1880s. At the time, Edison Electric dominated the market with its direct current (DC) power systems. One of the chief advantages of AC systems, which are the dominant source of power today, is that the electricity can be transmitted over longer distances from the generation source. The early DC systems were introduced in densely populated urban areas such as Manhattan, where distance was less important. Recognizing the potential threat from Westinghouse, Edison launched a major campaign to discredit the AC systems by charging that they were dangerous to consumers. Thomas Edison used his household name as a springboard for spreading the word, pointing out, for example, that the Westinghouse system was used for electric-chair executions. Westinghouse was forced to respond with a bold demonstration to prove its system was safe.

Enter Nikola Tesla, the famous physicist (born in Croatia to Serbian parents) who had once worked for Edison and was now an executive at Westinghouse. Tesla devised a demonstration that many people today have seen in a high school physics class or museum. A portable dynamo is used to generate an electrical current, and a volunteer is asked to place a hand on the device and conduct the current. Even though the subject's hair literally stands on end from the current, he or she remains completely unharmed. This demonstration was used as compelling proof that AC electric technology was a safe source of household energy.[15]

One of the most promising technologies of the future is that of genetically modified crops. To date, at least 10 varieties have been approved for sale in the United States, from papayas to potatoes. A third of the corn crop and half of the soybean crop is grown from genetically modified seed, which benefits farmers by offering programmed defenses against pests, producing higher yields and drastically lowering the cost of production. In the future, this technology may result in wonder foods that lower cholesterol, boost immunity, and provide more nutrition.

Yet agribusiness in the United States did not heed the human tendency toward insecurity about technology, resulting in a marketing catastrophe that is likely to constrain the growth of this technology for years to come. American farmers in large numbers were encouraged to plant genetically modified strains in the late 1990s, and major food companies relied on them for ingredients. Meantime, there were ominous rumblings, particularly in Europe, in the form of consumer protests and scientific criticism about the unproved safety of these products to consumers and of their potential harm to the environment. Agribusiness did not pay sufficient attention to these concerns, nor did it take adequate steps to reassure the market. Ultimately, the issue blew up in Europe as countries started to impose labeling laws, and retailers and food producers were forced to pull products with genetically modified crops off the market. Worse, the business and public relations disaster gained attention in the United States, raising concerns in a market that has traditionally been more receptive to agricultural innovations.[16]

The next revolution in innovation is likely to be based on biotechnology. The lesson to American agribusiness and other purveyors of these new technologies is to realize the extent of insecurity that consumers have about any innovation. Efforts need to be undertaken to "prove" the safety of these technologies, and the level of proof will no doubt have to be more than is acceptable to government regulators.

Reassuring Features and Messages

Many technology marketers rely on less dramatic forms of communication to reassure customers. One of the first applications of Muzak, the familiar background music provided by the Seattle company founded in the 1920s, was as a tool for technology reassurance. The technology was the high-rise elevator, and the music was meant to calm passengers nervous about their safety as they ascended floors.[17]

As one of the newest and most criticized technologies, Internet service providers take many steps to create a greater sense of security. This includes messages about how to protect against password theft, online police to ferret out problem users in a community, site areas and hot lines for reporting problems, and periodic messages about efforts to ensure safety. One of the biggest fears about the Internet is that it is unsafe for children. America Online offers the ability to activate parental controls on children's accounts, blocking access to problem sites and restricting uses that might be viewed as harmful. Other ISPs will direct their users to services such as "Net Nanny," which provide a level of protection from Internet smut.

Assuring Feedback

Technology needs to continually reassure the user that it is working as promised. CollegeNet is a national Web site that allows high school students to apply for colleges online. The site has many features that make this process easy. For example, users can apply to several colleges at a single Internet location. Information that the applicant enters on the site will be stored in a file that recalls the information whenever needed, reducing the drudgery of filling out forms. One of the greatest insecurities that any college applicant will have is wondering whether the application actually arrives at the admissions office. A feature of CollegeNet is a personal log where users can access a record of transactions, seeing what they submitted and what was received by colleges. This feature not only serves as a record-keeping convenience, but also provides a high degree of assurance that the process is working reliably.

The concept of reassuring feedback is used in a variety of technologies where customer insecurity is taken seriously. Automated gasoline pumps offer a receipt as an option and thank the customer when the transaction is complete. Users of e-tickets for airline travel are issued a piece of paper, not a true ticket but an assurance that the flight is booked and a seat awaits. In each case, the principle is the same: provide a reassuring form of feedback, realizing that the less-techno-ready consumers will not rest easy until they are certain the technology fulfilled its promise.

Brand Names

In a competitive environment, marketers obsess over brand equity and brand positioning, often forgetting an original historical purpose of branding a product or service. A solid brand name is a form of guarantee, a signal that the consumer is engaging in a risk-free purchase. The brand name can be a sign of a clean hotel room, unadulterated food, a reliable appliance, or a financially sound bank. When technology is involved, there is a high degree of uncertainty, and marketers need to send a signal that they are engaging in a safe transaction. The brand identity is a valuable tool for assurance.

Techno-ready marketers have two choices with respect to branding: create a strong brand name from scratch or leverage an existing brand name. Many of the new entrants in the dot-com world invest huge sums exclusively in creating a brand identity. An ad during the Super Bowl is considered as critical to corporate strategy as the technology itself. A widely recognized name conveys a natural sense of security. The brand is a recognized presence, so the consumer with a high level of insecurity is more willing to take a chance on the techno-product or service offered by a reputable brand. One of the major advantages of building a brand name from scratch is that it can be created in the image desired by the marketer. It may also be essential because the technology is so new; the idea of a search engine like Yahoo! has no precedent outside the online world.

A company should not, however, underestimate the value of an established, perhaps not so high-tech brand name. Amazon.com spent millions building a national brand presence. It must have been terribly frustrating when Barnes & Noble entered the e-commerce world with its own Web site. The established bookseller had a recognized name, and a consumer paranoid about who they are doing business with would find it reassuring to deal with this established company in an online environment.

We once consulted for a well-known Fortune 100 company that planned to vend a service on the Web. The firm made a deliberate decision to treat the service as a generic offering, perhaps to contribute to the impression that it would be inexpensive. Consumers who evaluated the concept were selected specifically because they tended *not* to be brand loyal. But when the consumers began to discuss the concept, they wanted to know *who* was offering the service. If it was not a recognized name, they would prefer not to use it unless they could do business by telephone rather than the Web. Brand loyalty normally meant nothing to these consumers, but when the service was delivered over a technology-driven medium, the company either had to rely on a more old-fashioned service model or reveal its identity to them.

Guarantees

The final option for a marketer is to offer a sound guarantee. Guarantees that protect the consumer from risk certainly help the cause of any techno-offering, but they are critical if the brand name is new or unfamiliar. eBay, an online trading community that matches buyers and sellers, helped create a new techno-service category by guaranteeing the integrity of its transactions. On an auction floor, a buyer would have the opportunity to inspect the merchandise, but in cyberspace, there is an unavoidable potential for fraud that poses an obstacle if not addressed. eBay offers its buyers free insurance, which covers up to $200 in losses in the event a seller fails to deliver the goods or provides less than the buyer expected. Thus, consumers are guaranteed by eBay that their transaction is safe, reassuring the more insecure consumers and generally making the whole concept of online auctions feasible.[18]

Any kind of guarantee is reassuring, and our countless conversations with consumers reveal that they like to see something in writing. It helps to tell them what happens if a high-tech appliance breaks, or an online order does not arrive by the holiday, or an unauthorized person gets access to account information and uses it to make purchases.

This chapter focused on one important principle: that satisfying customer needs is unique for a techno-offering. Unlike conventional products and services, those involving technology encounter resistance owing to an inherent discomfort and insecurity on the part of a large segment of the population. Marketers can address these concerns through customer-focused design, responsive customer care, and reassuring communication. In doing so, the technology becomes accessible to more than just the techno-elite and will reduce the burdens of cleaning up after failed attempts to use technology.

CHAPTER 9

The Techno-Ready
Marketing Audit

Technology businesses are faced with a unique set of decisions pertaining to their marketing planning process. They need to reexamine, and revise as necessary, their assumptions about how their markets operate, how their customers think, and which tactics are most effective in achieving growth. One approach such businesses can follow for planning effectively is to conduct a Techno-Ready Marketing Audit, consisting of a comprehensive review of the marketing environment, organization, and strategy. This process is similar to one a marketer of a traditional product or service might undertake, but it takes into account the special circumstances involved for a cutting-edge technology.[1] A marketing manager can use an audit framework as a checklist to identify gaps in strategy and ideas for ensuring success. Or an audit framework can be used to guide a thorough marketing planning process in a techno-driven environment.

The Techno-Ready Marketing Audit consists of a series of diagnostic questions that begin as highly strategic and long-term in focus and become increasingly tactical and immediate. We have grouped these issues into five major categories, as shown in Exhibit 9–1 and outlined below:

- Macro-Environment: the long-term trends and environmental variables that should influence the marketing organization and strategy

- Customer: the characteristics of prospects and of existing customers, and the structure of the market for the technology

- Marketing Organization: the techno-readiness of the management and employees, available resources, and the way they are organized to deal with a technology environment

- Acquisition Strategy: the program for obtaining new customers in an expanding and evolving technology market

- Retention Strategy: the program for supporting and satisfying customers of the techno-product or service

In the following discussion, we offer a checklist of

EXHIBIT 9-1

TECHNOLOGY MARKETING AUDIT AREAS

Macro-Environment

Customer

Organization

Acquisition
Strategy

Customer
Satisfaction
and
Retention
Strategy

issues and questions for each of the above categories. The checklists draw from and build on the various insights in the earlier chapters.

THE TECHNOLOGY MACRO-ENVIRONMENT

Marketers should conduct a broad survey of the environment for macro-trends when formulating the strategy for introducing their technology, focusing on trends in the development of their core technology as well as other technologies that might pose a competitive threat. There should also be a review of the regulatory environment, especially in light of an increasing tendency to enforce antitrust laws in the United States and the European Community.

The following is a checklist of planning questions related to the macro-environment:

▣ Are there opportunities, given the structure of the market, for our company or a competitor to achieve a critical mass that allows market dominance? This might occur as a result of the following:

- Brand equity: recognition as the premier provider in an industry
- Interconnectivity: a consumer benefit that increases proportionately as a result of linking increasing numbers of users to a system, pushing customers toward a single network or community
- Control of a revenue opportunity: the ability to derive substantial revenues because of a unique market position
- Control of a standard: a characteristic that makes the technology more valuable if all users adhere to a common standard

▣ Who are our competitors? Who offers a similar technology, and how is it differentiated in terms of features and applications? Who competes with us using an entirely different technology, and what advantages and disadvantages does that technology offer over our own?

☐ Which technology features are in development in the market that could enhance our current offering and allow us to provide more value to customers?

☐ Which technologies are in development that could potentially leapfrog our existing technology? Are there technologies or technology features that should be incorporated into future versions of our offering to ensure we are future-ready? Should we be aware of and plan for new technologies that would render our current offerings obsolete?

☐ At which stage of maturity is the market for the core technology upon which our product/service is based? Is it in the introductory, accelerating-growth, peak-growth, or declining- growth stage?

☐ Which important trends can be expected in the regulatory and legislative environment? Specifically:

- Are there important laws or rulings that will affect our technology?
- What is the position of governments regarding the imposition and control of standards?
- What is the impact of our technology on interstate and global commerce, and how are governments reacting to this in areas such as taxation?

- Is there a current or future threat of antitrust activities to our company, an important competitor, or an important partner?

▣ What view do important factions hold regarding our technology? Are there important ethical considerations, as perceived by our own stakeholders or different parts of society, that should be heeded in formulating communications and rollout of our technology?

THE CUSTOMER

Marketers should also take into account the consumer behavior process for their technology, relying on research to understand the level of consumer techno-readiness, the decision process for the technology, and perceptions of the technology from the consumer's standpoint.

Planning questions related to the customer include the following:

▣ What is the level of techno-readiness of different groups of customers? Specifically, what is the level of *optimism, innovativeness, discomfort,* and *insecurity* of

- Current customers who need to be serviced?
- The newest customers who need to be sold?
- Future customers whose needs our long-range planning should address?

▣ To what extent is the market for the technology

dependent upon replacement purchasers and customers who are expanding the number of units they own? How does the techno-readiness of these consumers differ from that of the market as a whole?

◨ Do particular aspects of techno-readiness predominate in our market? Specifically, does the challenge consist of clarifying benefits, leveraging innovators, making technology easy to use, or allaying fears? Or, do we face a combination of these challenges?

◨ From the customer point of view, what are the major benefits of our technology and how do they affect the lives of users? What language and symbols are most effective in communicating these benefits?

◨ What is the financial "business case" that can be made for buying our technology? What is the bottom-line impact on the income and expenses of a household or business from acquiring our technology?

◨ What is the decision process for acquiring the technology? Who are the decision-makers and influencers within a household or organization? Are there key individuals high in techno-readiness who sway the purchase?

THE TECHNO-READY
MARKETING ORGANIZATION

Companies should examine their own organization and its ability to market and support a techno-driven offering. This review includes a look at the technology readiness of all organizational levels, resources for marketing, and systems for supporting employees who work with customers. Key checklist areas for the organization include the following:

▣ What is the level of techno-readiness, or receptivity to new technology, among different players in the organization? Specifically, how techno-ready are

- Senior managers who set the overall technology direction for the organization?
- Midlevel management who carry out technology marketing strategy?
- Line employees and customer care staff who interact with customers?

▣ There has to be a balance in the level of techno-readiness, so the following questions need to be asked:

- Are there gaps in views on technology between customers and employees, customers and management, management and employees?
- Are any groups too techno-ready, such that they fail to comprehend the daily concerns of the average customer?
- Are any not techno-ready enough, so that they fail to share the vision of Explorers and Pioneers?

▣ Does the organization market low-tech products/ services as well as cutting-edge technology? If so, is there proper coordination and integration between the high-tech and low-tech areas? Are the different areas in alignment in their overall marketing strategy, or do they operate in a manner that creates conflict within the organization or in the marketplace?

- ▣ Is technology used internally in a manner that enables the organization to fulfill its marketing mission? The following questions should be posed:

  - Do information systems provide linkage between employees and customers (e.g., can an employee in customer service access information from an e-commerce system in order to solve a customer problem?)? Does technology enable employees—that is, provide power and control—to effectively serve customers?
  - Is technology leveraged to give management superior control over marketing activities? Do customer information systems link management to customer feedback?
  - Do systems provide managers with information on market trends and marketing performance?
  - Do systems link management to the organization, helping to create consistent messages and gather feedback?

- ▣ Does the organization have a rigorous market research process for developing and evaluating technology? Specific areas to address include the following:

  - Is there a formal process for designing and

refining techno-driven products/services based on customer feedback? Are highly techno-ready consumers targeted for their specific feedback, to ensure that the technology is future-ready?

- Is there a process for conducting usability testing of our techno-driven products/services? Are low- to medium-techno-ready consumers targeted for their specific feedback to ensure that the technology is customer focused for all users?

- Is there a process for identifying next-generation technologies and applications and disseminating information to technology designers who need to respond?

▣ Is there a strong teamwork orientation between technology designers (engineers, information systems professionals) and staff with more specialized marketing roles?

ACQUISITION STRATEGY

A firm's acquisition strategy—the process of obtaining new customers—is critical for techno-driven products and services because they are being introduced in emerging markets where the bulk of the opportunity resides in new customers. The techno-ready organization should formulate a marketing strategy that is predicated on the unique buying process for a technology. The decisions for a techno-product or service can be organized in the same fashion as for a traditional offering according to the 4 Ps of marketing: product, promotion, price, and place.

Product

☐ Are our products future-ready, meaning do they contain functionality and applications that are likely to be in demand in the near-term future? In the growth stages of the market, do our products meet the needs of the more techno-ready segments of the market?

☐ Is there a danger that our techno-offering is too advanced for the market? In other words, can we expect demand in a reasonable time frame, or is the time for the market to recognize the value of our offering so far off that a new generation of technology may displace it?

☐ Are our products and services customer-focused?

- Do customers find their use intuitive, efficient, responsive, and assuring?
- Are our products reliable or does more work need to be done to work out problems to ensure that they are suitable for all users, not just the techno-ready?
- Is our technology compatible with other technologies that the typical customer will use? Has this been verified through research that profiles technology ownership?

☐ Are our techno-driven products/services continually

updated and refined to meet the expanding needs of the more techno-ready consumers?

🔲 Do our techno-products/services offer sufficient capability for customization by the more techno-ready consumers? Are the "default" options designed to meet the needs of the less techno-ready consumers who want to avoid customizing?

🔲 As the market matures, are we providing more basic, simpler offerings to appeal to the less techno-ready consumers who have not yet purchased within our category?

Promotion/Communication

🔲 Do we have an established brand name that can be leveraged to promote and ensure confidence in our techno-product/service? Is the brand name suitable for a cutting-edge market, or is minor repositioning or an alliance with another company or brand needed to build confidence?

🔲 Is there a need to create a new brand name? Are adequate resources available to achieve a strong brand identity in the target market, and if so, is the investment worth it (e.g., is it cheaper to buy a recognized brand)?

🔲 Do we have a formal statement of the unique benefits of our techno-offering that is framed in the language of the consumer? Has this been tested among consumers to ensure that the promised benefits are meaningful to them?

☑ How do we communicate the benefits of our technology to the market? Are we using effective methods to demonstrate and clarify the benefits through traditional promotional channels (advertising, publicity, direct marketing, sales) and new media channels (Web sites, advertising)?

☑ Do we have a targeted message for highly techno-ready consumers? Do we have a process for technology evangelism, which could include paid professionals, in which we reach out to early adopters and opinion leaders (writers, technology workers, serious consumers)? Are opinion leaders provided with tools and information to talk about our techno-products/services?

☑ Have we adequately assessed the potential concerns and fears that might arise as a result of our technology? Are we tracking for emerging myths or real complaints about our technology's failure to deliver as promised in a safe fashion? What kind of communications program do we have in place to reassure the market?

Price

☑ What is our revenue model, or plan, for providing income from our products/services? Have we explored all avenues besides unit sales (e.g., selling advertising, customer support, sales of add-ons, sales of supplies)? Do we have the optimum model to ensure a competitive and appealing offering, and to achieve our long-range growth and retention goals?

▣ Are we continually measuring the price elasticity in the market and updating our assumptions as the market matures? What message does our price communicate about the quality and safety of our techno-products/services?

▣ Do we offer a range of pricing options to appeal to a wide range of consumers, from those who are sophisticated and demanding to those who are skeptical and hesitant to buy?

▣ Have we taken advantage of tactics for providing value for less techno-ready, more cost-conscious consumers, short of outright discounting? Examples of such tactics include offering basic stripped-down models, creative packaging and bundling, use of alternative sales channels, and temporary promotions.

Place, or Distribution Channels

▣ Do we take advantage of technology to distribute our offerings in a manner that is efficient and optimally convenient for the highly techno-ready consumers (e.g., distributing on the Web)? Have these distribution channels been tested to ensure that they are easy to use and as customer-focused as the core products and services?

▣ Do distribution channels offer the kind of support and interaction required by consumers with high levels of discomfort and insecurity with technology? Can buyers get access to a live representative by telephone? Is our technology sufficiently complex that it should be distributed or supported through in-person channels such as retail storefronts?

▣ Do distribution partners, such as retailers selling our products, share our vision of our technology? Can they support customers with comfort issues?

▣ How is our sales force managed to ensure a responsive interaction for a techno-offering? Specifically:

- Are sales staff interested in the technology, making them effective and enthusiastic missionaries? Are efforts made to recruit representatives who are above average in techno-readiness?
- Are sales staff trained to communicate effectively with low-TR consumers about product benefits? Do they possess technical knowledge to meet the information needs of high-TR consumers?
- Is technology used to enable and empower the sales force during customer interactions?

RETENTION STRATEGY

The long-term success for a techno-ready marketer hinges on the ability to satisfy and retain customers. For a techno-product, retention consists of ensuring repeat purchases, which come in the form of replacing units or expanding them within a household or business. Retention for a techno-service consists of continuing the relationship and, when the customer has the ability to use multiple services, capturing more of their total business. Retention depends on ensuring customer satisfaction, a task that is more challenging than for a low-tech product or service.

The following are key checkpoints to ensure that customer satisfaction is being managed in a manner suitable for a technology context:

- Is there a responsive customer care function that supports users with varying levels of techno-readiness? Some specific questions:

  - Are customers given an option for responsive, personal technical support? Is online, telephone or face-to-face support available as needed, depending on the technology?
  - Have adequate resources been allocated to customer care? Can customers get access at convenient times and in a reasonable time frame?
  - Are customer service representatives recruited and trained to ensure that they enjoy working with customers on technology problems? Do they have a versatile style that allows them to be able to interact with customers with varying levels of techno-readiness?
  - Do we regularly rely on our customer service representatives for feedback about our products?

▣ Are customers empowered to solve problems on their own? Do customers have access to sufficient options for self-help, including documentation?

▣ Is there a process for tracking customer satisfaction and providing feedback to the decision makers? Are the results analyzed by level of techno-readiness among customers? Are the products and services continually monitored, and modified when necessary to improve usability and reliability?

▣ What actions are taken to ensure safety and reassure customers? Are there adequate functions in place to protect customers? Are these safeguards communicated to customers?

▣ What kinds of guarantees are offered to protect against product failures, safety failures (e.g., stolen information), obsolescence?

▣ Are products and services marketed in a fashion that encourages long-term relationships (e.g., providing discounts for advance payments, recognizing relationships by offering rewards, protecting against obsolescence)?

IMPLEMENTING A
TECHNOLOGY AUDIT

In the fast-paced environment that technology marketers face, long-range strategic planning may seem an anachronism. In the span of a one-year planning horizon, a company can dominate an entire market, and in the span of a five-year plan, a company can grow into a behemoth like Microsoft or AOL. Nevertheless, it is important to institute some kind of conscious marketing planning process that formally addresses the realities of the technology market environment, studies the customers, examines the organization and its resources, and translates the findings into effective strategies for success.

The marketer of a technology must view the world through a special filter that recognizes the unique market dynamics and consumer processes that distinguish a cutting-edge technology from a traditional product or service. The Techno-Ready Marketing Audit provides such a filter, arming the technology marketer with a checklist of special considerations that must be addressed in order to be an effective marketer of innovation.

A final thought we offer is that the world at the start of the new millennium appears to be driven by increasing change, with innovation becoming the standard competitive strategy for companies. Nearly every decision faced by any marketer somehow involves something that is novel and cutting edge. For example, even if the product itself is low-tech, say, bath soap or cereal, the modes of distribution and promotion are likely to involve technology, say, e-commerce and new media. What does this mean for marketers in all sectors of the economy? The questions addressed in the Techno-Ready Marketing Audit become applicable to all of us involved in teaching, researching, and practicing marketing. Everyone in the future may need to begin seeing markets as a process of continuing innovation, technology diffusion, and changing patterns of consumer behavior. The bottom line is that all of us in marketing will need to grasp a new philosophy for effectively practicing the science of marketing.

CHAPTER 10

The Techno-Ready Society

This book—a discussion of theories of technology adoption and their application in marketing—focuses primarily on goods and services in a for-profit context. We intentionally confined the scope of the book to bring into bold relief the challenges of techno-ready marketing in the commercial sector. However, the ramifications of technology consumer behavior extend far beyond for-profit businesses. Our technology research, and its implications for effecting change, have applicability in other sectors of society, including:

- Nonprofit institutions, such as educational establishments and hospitals

- Professional associations

- Charities and cause-related organizations

- Governments

All of these organizations are, in effect, "social marketers" charged with promoting positive change in society. Just as sellers of traditional products and services are faced with opportunities and challenges that result from technological change, so do these nonprofits. And, in the process of leveraging technology to advance their causes, they must recognize the unique behavioral processes underlying the adoption of their programs and satisfaction with their progress.

The importance of technology in society is evident from two perspectives: the paths carved by nonprofits in using cutting-edge technology, and the image that citizens hold about technology. Consider the following examples:

- Educational institutions are leveraging information technologies for distance learning, delivering curriculum via videoconferencing and the Internet, and offering students flexibility and savings in time. According to the NTRS, 47% of American adults feel it would be desirable to attend a class or lecture online, where the student can send and receive information from the teacher and classmates electronically.

- Organizations are relying on the Internet as a tool for disseminating information. For example, a group involved in promoting land conservation uses the

Web as a medium for linking landowners with experts who can help them with conservation planning. An educational association uses the Web as a tool for interesting children in mathematics, and a professional association allows members to access research information electronically. According to the NTRS, 23% of the public have visited a Web site for a membership organization they belong to in a 12-month period. This is more than the share of the public who have engaged in e-commerce in the same time period.

- Technology is used as a medium for pledging to charities. In the United Way campaign, employees of some organizations can rely on telephone menus, while others can use an online medium, making the process efficient for employers and confidential for employees.

- Governments are also on the forefront of using technology to increase their efficiency and level of constituent service. Drivers in some states can renew their automobile registrations online. Commuters breeze past tollbooths equipped with sensors that automatically charge the toll to their credit cards. Researchers can download detailed demographic data from the Census Bureau Web site or pictures from the Library of Congress. According to the NTRS, in a 12-month period, 14% of the public have contacted a government agency or politician online.

Techno-ready marketing in a social and nonprofit context is worthy of its own separate book, but much of what we have covered here has application in these other areas. The underlying theory of technology readiness is universal and applies to all behaviors, not just the consumption of commercial goods and services. A nonprofit institution or government deploying innovative technologies would face the same drivers of and obstacles to success, the same issues of optimism, innovativeness, discomfort, and insecurity.

The importance of technology readiness is evident in the different levels of acceptance of technology in a not-for-profit context. For example, 52% of highly techno-ready consumers would find it desirable to vote in a referendum for local government from a computer, compared with only 31% of those low in techno-readiness. All the examples described above relating to online education, association Web sites, and accessing government have greater acceptance among the more techno-ready citizens.

In the future, technology promises to do great things to transform our world, but social marketers should pay no less attention than their business counterparts to the unique consumer beliefs about technology. It is still important to demonstrate the benefits about technology used by nonprofits, to tap and mobilize innovators in advocating a new technology in society, to make technology designed for constituents customer-focused, and to expect and address public fears about technology.

The nonprofit and government sectors must deal with a special kind of paranoia, such as the following beliefs among the public found in the NTRS:

- 73% of the public believe that new technology makes it too easy for governments and companies to spy on people.

- 61% of the public believe that technological innovations always seem to hurt a lot of people by making their skills obsolete.

Technology holds much promise for our world in the future, streamlining government, promoting positive ideas, advancing public health, increasing the living standard of the poor, and the like. However, to bring the promise of technology to fruition, public-service organizations must be sensitive to the natural resistance toward cutting-edge technology and the technology-related concerns of their constituents.

Our years of research go much further than explaining the competent producer and servicer. They shed light on a pervasive theory of behavior that encompasses all facets of our lives where technology is involved. The paradox of technology beliefs—the inevitable love-hate relationship we have as consumers and as citizens—is an issue for all of us as a society as we play host to a new age of innovation.

Notes

Chapter 1

1. An excellent source of information on Thomas Edison and nineteenth-century technology is a visit to the Edison National Historic Site in West Orange, NJ, which is maintained by the National Park Service.
2. One company that did achieve the status of an industry leader was General Electric, which was formed in 1892 by consolidating Edison General Electric with other companies in the burgeoning electric industry. Edison broke away from GE in its early years when it became clear that he would not control the new company.
3. Neil Baldwin, *Edison: Inventing the Century* (New York: Hyperion, 1995), pp. 316–317.
4. Ibid., pp. 318–319.
5. Ibid., pp. 397–398.
6. The Greek word *techno* means a body of knowledge; the Latin word *novalisis* refers to a field plowed for the first time.
7. Source: AOL Communications.
8. The National Technology Readiness Survey Research report, May 14, 1999, published by Rockbridge Associates, Inc., Great Falls, Va. The 2000 NTRS reports that AOL's consumer market share had dropped to 35%.

Chapter 2

1. See Train World, The Golden Age of Railroad, www.stormloader.com/ironhorse/golden.html. Tourists interested in the story can visit the actual site of the race and see a replica of the Tom Thumb at the B&O Railroad Museum in Baltimore.

2. C. H. B. Quennell and M. Quennell, *Everyday Life in Prehistoric Times* (New York: G. P. Putnam's Sons 1963), p. 149.
3. Elizabeth L. Eisenstein, *The Printing Press as an Agent of Change,* Vol. 2 (Cambridge: Cambridge University Press, 1979), p. 507. One interesting criticism of printing is that it also inhibited innovation by preserving old knowledge.
4. See http://www.nao.otis.com/aboutotis/companyinfo/elevators.html.
5. Neil Baldwin, *Edison: Inventing the Century* (New York: Hyperion, 1995).
6. *The National Technology Readiness Survey, Research Report,* May 14, 1999, published by Rockbridge Associates, Inc., Great Falls, Va.
7. 38% at the beginning of 1999.

CHAPTER 3

1. A. Parasuraman, "Technology Readiness Index (TRI): A Multiple-Item Scale to Measure Readiness to Embrace New Technologies," *Journal of Service Research,* May 2000, pp. 307–320.
2. David Glenn Mick and Susan Fournier, "Paradoxes of Technology: Consumer Cognizance, Emotions, and Coping Strategies," *Marketing Science Institute Monograph,* Report No. 98–112, 1998.
3. Scott Hogenson, *Porn-Blocking Software Has Low Profile,* Conservative News Service, May 19,1999, www.conservativenews.org/culture/archive.
4. Peter McGrath, "Knowing You All Too Well," newsweek.com/nw-srv/issue/13_99a.
5. Social Effects of the Internet—Information Overload," http://polaris.umuc.edu/~dhenders/social_work.html.
6. "Social Effects of the Internet—Access to More [Unreliable?] Resources," http://polaris.umuc.edu/~dhenders/ social_ work. html.
7. "Social Effects of the Internet—Effect on Human Interaction," http://polaris.umuc.edu/~dhenders/social_problems.html.
8. "Social Effects of the Internet—The Cost of Inexperience," http://polaris.umuc.edu/~dhenders/social_problems. html.
9. "Social Effects of the Internet—Diminishing of Nationalism

and Cutural Identity," http://polaris.umuc.edu/~dhenders/social_work.html.

10. Christopher S. Wren, "A Seductive Drug Culture Flourishes on the Internet," *The New York Times,* June 20, 1997, p. A1; William Glaberson, "A Star on Your Computer Screen," *The New York Times,* December 25, 1997, p. A2.

11. Deborah Cowles, "Consumer Perceptions of Interactive Media," *Journal of Broadcasting and Electronic Media,* Winter 1989, pp. 83–89; Deborah Cowles, "Retail Bankers: Marketing May Not Be Enough," *Bank Marketing,* January 1991, pp. 22–25; Deborah Cowles and Lawrence A. Crosby, "Consumer Acceptance of Interactive Media in Service Marketing Encounters," *Service Industries Journal,* July 1990, pp. 521–540.

12. Pratibha A. Dabholkar, "Consumer Evaluations of New Technology-Based Self-Service Options: An Investigation of Alternative Models of Service Quality," *International Journal of Research in Marketing,* 13, no. 1, 1996, pp. 29–51.

13. The full 36-item TRI battery is copyrighted by the authors. For further information please contact the authors.

14. See note 1.

15. Details of these analyses are documented in the article referenced in note 1.

CHAPTER 4

1. Based on self-reported acquisition dates in the National Technology Readiness Survey.

2. "800,000 iMacs Sold in First 139 Days," Staff, ComputerCurrents.com, January 5, 1999, www.currents.net/newstoday/99/01/05/news11.html.

3. Nancy Weil, "Gateway Rethinks Consumer Strategy," *Computer World,* online news story, May 27, 1998.

CHAPTER 6

1. Christian Grönroos, "Relationship Marketing Logic," *Asia-Australia Marketing Journal,* 4, no. 1, 1996, pp. 7–18; Christian Grönroos, "Marketing Services: The Case of a Missing Product," *Journal of Business & Industrial Marketing,* 13, no. 4/5, 1998, pp. 322–338; Philip Kotler, *Marketing Manage-*

ment: Analysis, Planning, Implementation, and Control, 8th ed., (Englewood Cliffs, NJ: Prentice Hall, 1994, p. 470).

2. The pyramid model was first introduced in the following presentation: A. Parasuraman, "Understanding and Leveraging the Role of Customer Service in External, Interactive and Internal Marketing," presented at the 1996 Frontiers in Service Conference, Nashville, TN.

3. *The National Technology Readiness Study Report on Real Estate Agents,* presented to Interealty and participating multiple-listing services by Rockbridge Associates, Inc., July 1999.

4. Leonard L. Berry, "How to Sell New Services," *American Demographics,* October 1989, pp. 42–43.

5. Robert A. Ferchat (Chairman and CEO, Bell Mobility Cellular), "Avoiding the Iceberg in a Multi-Lettuce Age: Rethinking Customer Intelligence and Service Relationships," presented at the Sixth Annual Frontiers in Services Conference, Vanderbilt University, Nashville, TN, October 4, 1996.

CHAPTER 7

1. "The Rise and Fall of Apple, Inc., Part 2," *Rolling Stone,* April 18, 1996, pp. 59–63, 85–88.

2. "Apple Computer Apostles Keep the Faith and Preach the Gospel," Knight-Ridder/*Tribune Business News,* March 28, 1997, p. 328, from the *San Jose Mercury News,* Cal.

3. An example is www.TidBITS.com.

4. Sun Microsystems' Web site listed 34 technology and product experts, most with the title "Technology Evangelist," who were speakers during Sun Technology Days in January 2000.

5. Irvin M. May, Jr., "Agricultural Research and Education in Northeast Texas," Texas A&M University System, Overton, overton.tamu.edu/htmsub/history.html.

6. Frederick C. Fliegel and Joseph E. Kivlin, "Attributes of Innovations as Factors in Diffusion," *American Journal of Sociology,* 72, no. 3, November 1966, p. 235–248.

7. "Divx: The Video Technology that Geeks Love to Hate," *Fortune,* June 21, 1999, p. 177.

8. John Desmond, "Marketing Blunders Sideline Early Mainframe Vendors (Part 4)," *Computerworld,* May 27, 1985, p. 71.

9. Martin A. Armstrong, "The Anatomy of a Crisis," Princeton Economic Institute, http://www.pei-intl.com/Research/PANICS/INTRO.HTM.

10. "Coming of Age at Lotus: Software's Child Prodigy Grows Up," *Business Week*, February 25, 1985, p. 100.

11. Peter W. Bernstein, "Polaroid Struggles to Get Back in Focus," *Fortune*, April 7, 1980, p. 66.

12. Neil Baldwin, *Edison: Inventing the Century* (New York; Hyperion, 1995), p. 337–339.

12. Ibid., pp 318–319.

13. Brian Till, "Top 10," *Grand Connections*, St. Louis University, April 1998, http://www.slu.edu/publications/gc/v4-9/top_ten.shtml.

14. http://www.curcuitcity.com.

15. Public Debut (Corporate Report Research: The Public 200), Liz Brisset, Corporte Report-Minnesota, April 1999, p. 30.

16. "Select Comfort Corporation Announces Change in Strategic Direction for 2000," *Business Wire*, December 23, 1999.

17. "New Advantix Campaign Educates Users about APS," *Drug Store News*, May 19, 1997, p. 25.

18. The net present value refers to discounting the cash flow of benefits received over time. For example, if the dollar benefits in one year are $1,000,000 and the expected rate of return on investments is 10%, these benefits would be discounted to $909,091 to reflect the lower value relative to $1,000,000 today. Another way of viewing the net benefit is to divide the long-term benefit over the initial investment costs to calculate a return on investment.

19. The net benefit may also include "intangible" costs, which are hard to quantify objectively, such as avoiding management and employee stress during the peak season. Management would need to assign a dollar value using judgment, a process that could be aided by a structured survey approach such as conjoint analysis.

20. Katherine Morrall, "IQ Tests for Smart Cards," *Bank Marketing*, March 1997, pp. 19–25.

21. Marjorie Sorge, "Ford Model T Is Car of the Century," *Automotive Industries/AI-Online.com*, December 20, 1999, http://www.ai-online.com/news/122099.

CHAPTER 8

1. We once administered a TR test to a group of IT professionals in an intelligence agency, and they turned out to be highly insecure; in their jobs, it may be smart to be overly concerned, since the downside risk of a breech can be much more costly than in a more commercial environment.

2. A. Parasuraman and Charles L. Colby, "Perceived Quality of Techno-Based Services: The Impact of Customers' Technology Readiness," presented at the Sixth International Quis Conference, Norwalk, CT, July 2, 1998.

3. Developed by A. Parasuraman, Valarie Zeithaml, and Leonard Berry.

4. Patrick Gerland, United Nations Statistical Division, "Software Development: Past, Present and Future Trends and Tools," presented at the NIDI/IUSSP Expert Meeting on Demographic Software and Micro-computing, Strategies for the Future, The Hague, June 29–July 3, 1992.

5. Barbara Kantrowitz, "At Last, a Really Big Mac," *Newsweek,* September 28, 1998, p. 85.

6. Mark Leibovich, "The Iridium Lesson," *The Washington Post,* May 24, 1999, pp. 12–14.

7. An interesting visit is to the Samuel F. B. Morse home in Poughkeepsie, NY. Morse's home could be considered the first "wired home" in history, including telegraph lines that run up to the sleeping areas. For a description of the development of the original Morse code, see Neal McEwen, "Did Samuel F. B. Morse Invent the Code as We Know It Today?" The Telegraph Office, http://fohnix.metronet.com/ ~nmcewen/vail.html.

8. See "Why Qwerty Was Invented," home.earthlink.net/~dcrehr/ whyqwert.html. It is a myth that the QWERTY design was intended to slow down fast typists. Ironically, the original purpose of Sholes's design, to avoid jamming by separating common letters, is now obsolete. Attempts have been made to develop a superior keyboard, including one by August Dvorak of Washington State University in 1932; while Dvorak's keyboard speeds up typing by making common letters more accessible, a study by the U.S. General Services Administration in 1953 found that the type of keyboard made no difference, since speed is determined primarily by the skill of the typist.

9. See http://www.att.com/technology/history/chronology/64 touch.html.

10. Alan Goldstein, "Gateway Gets Closer to Buyers," *Dallas Morning News,* January 4, 1999.

11. See www.erols.com.

12. See www.newcorp.com.

13. Lance Rosenzweig, "eCommerce Customer Care: Building Customer Loyalty On Line," PeopleSupport.com, presented at the ninth Annual Frontiers in Services Conference, Nashville, TN, October 21–23, 1999.

14. David Ignatius, "Beware the Private Cyber Snoops," *The Washington Post,* January 30, 2000, p. B-7. The article quotes a former spy-turned-cyber-snoop as saying, "If you're on the neighborhood ring [cable loop], you can put a sniffer on the cable and watch everything I do on my computer—stock trades, passwords, e-mails, everything."

15. Neil Baldwin, *Edison: Inventing the Century* (New York: Hyperion, 1995), pp. 200–202; the Web site for the Boston Museum of Science, http://www.mos.org/sln/toe/tesla.html.

16. Justin Gillis, "New Seed Planted in Genetic Flap," *The Washington Post,* February 6, 2000, p. H-1.

17. See http://www.muzak.com.

18. See http://www.ebay.com. The company also maintains a staff group called SafeHarbor that investigates and resolves problems such as fraud.

CHAPTER 9

1. For a succinct discussion of a traditional marketing audit, see Philip Kotler, *Marketing Management,* Millennium ed., (Upper Saddle River, NJ: Prentice Hall, 2000), pp. 708–711.

INDEX

A. Parasuraman is the James W. McLamore Professor in Marketing (endowed by the Burger King Corporation) at the University of Miami. He has done extensive research and writing in the areas of service-quality measurement and improvement, as well as the role of technology in marketing to and serving customers. In 1988, he was selected as one of the "Ten Most Influential Figures in Quality" by the editorial board of *The Quality Review,* co-published by the American Quality Foundation and the American Society for Quality Control. He has received many distinguished teaching and research awards. In 1998, Dr. Parasuraman was honored with the American Marketing Association's Career Contributions to the Services Discipline Award. He has published numerous articles in leading journals. Dr. Parasuraman is the author of *Marketing Research,* a college textbook, and coauthor of two other business books published by the Free Press: *Delivering Quality Service* and *Marketing Services.* He is an active consultant for major corporations and has conducted scores of executive seminars in many countries. Dr. Parasuraman is a Senior Fellow of the University of Maryland's Center for E-Service.

Charles Colby is the founder and president of Rockbridge Associates, Inc., a market research firm based in the Washington, D.C., area that specializes in technology issues for services and information industries. He has provided strategic consulting and research to marketers for over twenty years, working with Fortune 500 companies, midsize growth firms, and nonprofits, all sharing a common interest in using technology to package and deliver their services. Prior to founding Rockbridge in 1992, he worked for Citicorp, Opinion Research Corporation, and Westat, Inc. He has written numerous articles on technology marketing and service quality, and has presented extensively to financial services, communications, and marketing management forums. He has an M.B.A. from the University of Maryland at College Park. His clients include MCI WorldCom, Discovery Channel, Verizon, Marriott International, Capital One, Freddie Mac, Cox Interactive Media, Best Software, and many professional and trade associations. Mr. Colby is a Senior Fellow of the University of Maryland's Center for E-Service.